# SpringerBriefs in Well-Being and Quality of Life Research

More information about this series at http://www.springer.com/series/10150

Seung Jong Lee · Yunji Kim
Rhonda Phillips
Editors

# Community Well-Being and Community Development

## Conceptions and Applications

Springer

*Editors*
Seung Jong Lee
Seoul National University
Seoul
Korea, Republic of (South Korea)

Rhonda Phillips
Purdue University
West Lafayette, IN
USA

Yunji Kim
Cornell University
Ithaca, NY
USA

ISSN 2211-7644 ISSN 2211-7652 (electronic)
ISBN 978-3-319-12420-9 ISBN 978-3-319-12421-6 (eBook)
DOI 10.1007/978-3-319-12421-6

Library of Congress Control Number: 2014953310

Springer Cham Heidelberg New York Dordrecht London



Printed on acid-free paper

Springer is part of Springer Science+Business Media (www.springer.com)

# Preface

The idea for this volume began when a group of scholars, the Community Well-being Research Network, convened in 2012 at Seoul National University. With representatives from throughout Asia, Europe, and the US, ideas began to coalesce around community well-being as applied to community development and societal well-being. This collection of chapters represents the outgrowth of that convening, along with another volume in the *SpringerBriefs* series, *Learning and Community Approaches for Promoting Well-Being* edited by Youngwha Kee, Yunji Kim, and Rhonda Phillips. It is our intent to spur interest in community well-being with both conceptual and applied work.

Several perspectives on community well-being and its relationships to community development are provided. Both these areas are related and highly interlinked. This collection of four chapters provides exploration of the underlying concepts and foundations as well as applied case studies illustrating the connections between community development and community well-being.

The first chapter, "Exploring the Intersection of Community Well-Being and Community Development," by the editors discusses the relationships between these two areas and how one impacts the other. Differences as well as commonalities are explored. Several contexts are provided for promoting understanding of the interconnections across these areas of scholarship and practice.

The next chapter, "Searching for the Meaning of Community Well-Being" by Seung Jong Lee and Yunji Kim, predominantly focuses on concepts of community well-being. It presents a framework for considering community well-being as an encompassing concept, touching on dimensions of quality of life, happiness, sustainability, and other community concerns. Additionally, "Searching for the Meaning of Community Well-Being" provides a brief history of the concepts of community well-being from its basis in ancient times to current time.

The third chapter, "Building Community Well-Being Across Sectors with "For Benefit" Community Business" by Rhonda Phillips, discusses community development from a different vantage point. Looking at community-focused and community-owned businesses, this chapter explores ideas for strengthening local economies via methods and policies to support local business development.

It presents policy suggestions for developing local ownerships programs, and relates these activities to tools and techniques to aid in progress toward promotion of overall community well-being and development. This chapter also seeks to provide illustration of an alternative way of thinking about these concerns related to economic and social well-being, from the vantage point of community well-being and community development.

The volume concludes with the second case study in the last chapter, "Community Bonding and Community Well-Being, Perspective from a Community Development Council in Singapore," by Leng Leng Thang, Seung Jong Lee, and Youngwha Kee. This chapter discusses the loss of sense of community due to rapid urbanization. Factors such as economic, socio-cultural, and government policies have played a role leading to the demise of community sentiments and attachment. Urban sprawl as a result of government public housing and new town policies have uprooted residents and disrupted pre-existing communities. In Singapore, public housing policies in mass scale—while well recognized for its effectiveness in meeting serious housing shortage—are also said to have caused the loss of community from the prior kampongs (village) style of living and sense of place. This chapter presents strategies and policies for helping to create a greater sense of community and community development outcomes, as desirable outcomes for fostering improved community well-being.

Our purpose in compiling this volume is to promote more scholarship and application at the intersection of community development and community well-being. The potential benefits of more closely aligning these two areas holds much promise for our communities.

Seung Jong Lee
Yunji Kim
Rhonda Phillips

# Acknowledgment

This work was supported by the National Research Foundation of Korea Grant funded by the Korean Government (NRF-2013S1A3A2054622).

# Contents

# About the Contributors

**Youngwha Kee** is a professor in the Department of Lifelong Education at Soongsil University and president of the National Institute of Lifelong Education. She received her Ph.D. in Education and her Master's in Public Administration. She currently serves as the director of the Korea Institute of Local Development Education. Previously, Dr. Kee was president of the Association of Adult and Continuing Education of Korea and researcher of Korea Association for Community Education. Her book "Critical Theories for Adult Learning" (Translated) was selected as an outstanding academic book by the Korean National Academy of Sciences in 2010 and in 2007 she received the outstanding educator award by the Seoul Metropolitan City Human Resource Development Center. In Korea, she has served on several advisory committees in relation to educational policies and has been deeply involved with community education among multicultural families and the church leadership education center. Internationally, she serves on the editorial board of the *International Journal of Continuing Education* and *Lifelong Learning* (Hong Kong) and the *Lifelong Education Magazine* (Taiwan) and *Lifelong Education* (China). Her research interests include older adult learning, community education, civic education, community development, and governance.

**Yunji Kim** is a doctoral student in the Department of City and Regional Planning at Cornell University. She received her master's degree from the Graduate School of Public Administration, Seoul National University in 2011. Her master's thesis, "Development and Application of a Community Well-being Index for Korea," focused on defining community well-being as a guiding principle for local governments and suggested a new community well-being index for Korean metropolitan districts. Her current research interests include the relationship between community well-being and local government services; citizen participation; and community development. Her recent articles include, "The Development and Application of a Community Wellbeing Index in Korean Metropolitan Cities," "An Analysis of the Relative Importance of Components in Measuring Community Wellbeing: Perspectives of Citizens, Public Officials, and Experts," and "Sing, Dance, and Be Merry: a Strategy for Successful Urban Development?"

**Seung Jong Lee** is a professor at the Graduate School of Public Administration, Seoul National University and president of the Korea Research Institute for Local Administration. Previously, he served as the president of the Korean Association for Public Administration and chief editor of several academic journals in related fields, such as the *Journal of Korean Association for Policy Studies*, *Journal of Association for Local Government Studies*, and *The Korea Local Administration Review*. He has frequently advised local and national governments through such positions as chairman of the Local Government Administration Joint Evaluation Committee, vice chairman of the Presidential Committee on Local District Reorganization Plans, and member of the presidential transition committee. He has not only done extensive research on citizen participation and local autonomy, but has also been a strong advocate and educator in the field. He is the author of *Theories of Local Autonomy*, and *Democratic Politics and Citizen Participation*.

**Rhonda Phillips** Ph.D., AICP, is Dean of the Honors College at Purdue University and a professor in the Agricultural Economics Department. A former senior sustainability scientist in the Global Institute of Sustainability for Arizona State University, she served as professor in School of Community Resources and Development, and affiliate professor in the School of Geographical Sciences and Urban Planning. Community well-being and development comprise the focus of Rhonda's research and outreach activities including community-based education and research initiatives for enhancing quality of life. Honors include serving as the 2006 UK Ulster Policy Fellow Fulbright Scholar and a 2012 Fulbright Senior Specialist to Panama. Rhonda is author or editor of 18 books, including *Community Development Indicators Measuring Systems*, and *Introduction to Community Development,* and serves as editor for the Springer series, *Community Quality of Life and Well-Being*. Rhonda is president of the International Society for Quality-of-Life Studies, www.isqols.org.

**Leng Leng Thang** is an associate professor at the Department of Japanese Studies, the National University of Singapore. She is a socio-cultural anthropologist with an area focus on Japan as well as Singapore. Her research interests include aging, intergenerational programming and relationships, gender, and family. One of her most recent book publications is an edited volume titled "Experiencing Grandparenthood: An Asian Perspective" (co-edited with Kalyani Mehta, Springer Publishing, 2012). She is associate editor of the *Journal of Intergenerational Relationships* published by Taylor and Francis.

# Exploring the Intersection of Community Well-Being and Community Development

**Seung Jong Lee, Yunji Kim and Rhonda Phillips**

## Introduction

Community well-being is a term that varies in meaning by culture, group, society, and communities. Despite these variances, common frameworks of concepts and measures can assist communities in prioritizing goals and values. In this context, community well-being can be thought of as encompassing "the broad range of economic, social, environmental, cultural and governance goals and priorities identified as of greatest importance by a particular community, population group or society" (Cox et al. 2010: 72). Further, "the concept of community well-being is focused on understanding the contribution of a community in maintaining itself and fulfilling the various needs of local residents" (Haworth and Hart 2007: 95). As a means of social relationship and social organization, well-being is positioned as "something that we do together, not something that we each possess" (p. 128).

As noted in chapter "Searching for the Meaning of Community Well-Being" of this volume, a number of scholars use the word community well-being synonymously with socioeconomic factors. Given the rise of interest in community well-being and the range of potential applications, as well as development of underlying concepts and theoretical foundations, the concept is now considered to be more comprehensive than socioeconomics. While this remains a central and key feature of community well-being, other domains are increasingly being considered.

S.J. Lee (✉)
Seoul National University, Seoul, Republic of Korea
e-mail: slee@snu.ac.kr

Y. Kim
Cornell University, Ithaca, NY, USA
e-mail: yk634@cornell.edu

R. Phillips
Purdue University, West Lafayette, IN, USA
e-mail: rphillips@purdue.edu



S.J. Lee et al. (eds.), *Community Well-Being and Community Development*,
SpringerBriefs in Well-Being and Quality of Life Research,
DOI 10.1007/978-3-319-12421-6_1

For example, words often used in conjunction with community well-being include happiness, social quality, socioeconomic quality, quality of life, healthy community, and sustainability, leading to much discussion and debate about definition. To further expand the situation, community development has long addressed issues of well-being.

More comprehensive or holistic definitions include a number of community factors, such as physical, psychological, political, social, cultural, and environmental. Most cities and regions define community well-being as incorporating economic, social, and physical dimensions, particularly in the context of specific population groups (the "well-being" of children, for example). Cox et al. (2010) provides an integrated approach through encompassing economic, social, environmental, cultural as well as governance goals and priorities. CWB is both abstract and multidimensional, as noted in the following definition by Chanan (2002):

> [community] [is] a number of people who have some degree of common identity or concerns often related to a particular locality or conditions … a community is not a thing. It is a number of people who have repeated dealing with each other. When community is identifiable with a locality, CWB/the quality of community life is intimately connected with: how well that locality is functioning; how well that locality is governed; how the services in that locality are operating; and how safe, pleasant and rewarding it feels to live in that locality.

For this work, we define *community well-being as that state in which the needs and desires of a community are fulfilled.* We use this definition throughout the text and also to construct the analysis in the subsequent chapters.

## Community Development

Community development, both as a practice and a discipline, can be considered a complementary concept with overlapping core foundations as found in community well-being. The predominate focus and value of community development is to improve people's lives. It is a practice in the sense that many are involved at the local, regional and national levels from a public sector as well as from the nonprofit/nongovernmental sectors. In the US alone, it is estimated that many thousands of civic sector organizations are founded with a community development mission. Most governments around the globe have community development either explicitly or implicitly embedded within their policies, approaches and programs.

The Community Development Society, based in North America, and comprised of both academic researchers and community practitioners was formed in 1970—a direct result and outgrowth of the social movement coming to the forefront in the 1960s. Their Principles of Good Practice are as follows, and reflect the underlying values of community development as a practice, discipline and area of scholarship:

- Promote active and representative participation toward enabling all community members to meaningfully influence the decisions that affect their lives.

- Engage community members in learning about and understanding community issues, and the economic, social, environmental, political, psychological, and other impacts associated with alternative courses of action.
- Incorporate the diverse interests and cultures of the community in the community development process; and disengage from support of any effort that is likely to adversely affect the disadvantaged members of a community.
- Work actively to enhance the leadership capacity of community members, leaders, and groups within the community.
- Be open to using the full range of action strategies to work toward the long-term sustainability and well being of the community (Community Development Society 2014).

These principles show the mission and value-focused nature of community development, and many of these overlap with those underlying precepts of community well-being. Further, the International Association for Community Development, a nonprofit/nongovernmental association with members from around the world with the mission of sustainable community development for social justice, identifies the following as their priorities for fostering community development:

- Promote community development as a key method for addressing challenges, opportunities and priority issues in rural and urban areas locally, regionally and internationally.
- Facilitate quality practice exchange, education, training, research and publications in support of practitioners, educators, researchers, policy analysts, activists and other community workers and organizers.
- Engage practitioners, educators, researchers, policy analysts, activists and other community workers and organisers at country and regional levels and thereby promote their community-based planning and development work.
- Ensure the short, medium and longer term sustainability of IACD (IACD 2014).

A few more definitions are in order. Community development can be thought of as both a process and an outcome, as defined below:

> Community development is "a process of developing and enhancing the ability to act collectively and an outcome: (1) taking collective action and (2) the result of that action for improvement in a community in any or all realms: physical, environmental, cultural, social, political, economic, etc." (Phillips and Pittman 2009: 6).

It centers on capacity building and taking action. As seen, a vital part of the term community development is that of *community*. Let us reverse course for a moment and digress to exploring this word. Numerous definitions exist, although we will center on those that are more place-based in nature for purposes of this chapter. Definitions of community include the following:

1. People who live within a geographically defined area and who have social and psychological ties with each other and with the place where they live (Mattessich and Monsey 2004: 56).

2. …. “a general term to describe what occurs outside systems and institutions. It also refers to an aggregation of people or neighborhoods that have something in common. It is both a place and an experience of connectedness” (McKnight and Block 2010: 5).

As seen from these definitions, there is something special or connected about community, and has been for a very long time. The etymological roots are from the late 14th century, from the Old French *comunité* (translated as commonness or everyone. The Latin noun, *communitas* refers to “fellowship, friendly intercourse; courtesy, condescension, affability” (OED 2014). It also implies the spirit of community or *community spiritedness* with feelings of togetherness and/or solidarity uniting those in the community.

Connecting this special feeling of the common or togetherness with action to impart improvements *is* community development. There is a related area, that of community economic development, which is highly interrelated with community development itself although focused more on economic dimensions. We discuss it here because community well-being can be seen as the next generation of thought about measuring, gauging and guiding progress in societies as opposed to indicators such as the Gross Domestic Product. Efforts such as the Geniune Progress Indicator from Canada are more comprehensive although still include economic dimensions as a vital component in the now well known trilogy of economics, equity and environmental domains.

As noted, community economic development is highly interrelated with community development, with perhaps a more holistic approach to incorporating economic with other needs. Schaffer et al. (2006: 61) provide a description of this relationship:

> We maintain that community economic development occurs when people in a community analyze the economic conditions of our community, determining its economic needs and unfilled opportunities, deciding what can be done to improve economic conditions that community and then moving to achieve agreed upon economic goals and objectives.

Newer definitions incorporate vital concepts of sustainability and equity, reflecting the triology eluded to earlier. These include discussion of opening opportunities for the poor (Anglin 2011). Another definition is that offered by Phillips and Besser (2013: 6) with community economic development as a “merging of aspects of the fields of community development and economic development, implying practice aimed at community betterment and economic improvement at the local level, preferably encompassing sustainable development approaches.”

## Finding Common Ground

Community well-being encapsulates both community development and community economic development, as components fostering overall community well-being. The range of approaches, tools, techniques and policies in this arena vary widely, depending on local conditions and desires. However, there are a few areas of commonality. One area, due to the increase in interest in community well-being, is the development of community indicators to help ascertain conditions, gauge progress towards goals, and identify key trends some in the civic sector (public and nonprofit/non-governmental) and private sectors. These community indicators frameworks, sets and systems provide a basis of more relevant information to assist in decision-making and governance (Phillips 2003). When viewed as community well-being indicators, these systems become very useful in public policymaking, allowing residents to decide what quality of life means to them as well as other dimensions inherent in community well-being. With the development and use of community indicators, there is an emphasis on social inclusion and participation by community members; this serves development of "social trust and stronger communities via more meaningful and cooperative governance" (Rapley 2003: 45).

Community indicators are not new, they have been in use since the early 1900s in some countries as a way to gauge social progress. These social indicators provided insight into many changes in communities, regions and countries by tracking demographic changes, social trends and other valuable information.

Economic components are included in most, if not all, of these indicator sets, especially after World War II. These tend to be those typically accepted and used widely—per capita income, unemployment rates, numbers of jobs, and so on. However, just as with problems inherent in traditional measures of Gross Domestic Product, for example, these typical measures do not always fully reflect local values and goals. Issues of quality versus quantity are inherent with these type measures—will any job do in a community? What about the skills and assets in a community matching with the employment opportunities? How strong are the social aspects of types of business activities within a community? These and other concerns centering more on quality-of-life issues are increasingly being explored in communities.

*The Human Development Index*, pioneered by Mahbub Haq and Amartya Sen and used by the United Nations, is a commonly accepted model for gauging well-being across countries. A modified version, *The Measure of America*, uses similar standard measures calibrated for the US with the American Human Development Index. As with the Human Development Index, it incorporates measures that "reflect what most people believe are the basic ingredients of human well-being: a long and healthy life, access to knowledge, and a decent material standard of living" (Burd-Sharps et al. 2011: 115). These efforts and methods expand to larger domains and considerations impacting well-being beyond past measure.

One area that is not addressed enough is the idea of suffering in the collective (social suffering). Anderson (2014) explains that when others' suffering is alleviated or reduced, their quality of life improves. Justifying quality of life "as a concrete human need (with an) emphasis on *social* suffering as a qualitatively different type of suffering" (Anderson 2014: 22) fosters more appropriate and humane responses and perspectives. Connecting suffering alleviation to quality of life and overall community well-being will be an area of increasing interest and growth due to changing conditions around the globe.

Community indicators can then become a tool to help integrate considerations across the range of impacts and domains influencing well-being. Underlying all these discussions are foundations of the "collective" as well as concepts of equity, and improvement or progress in the human condition. Community indicator systems can then become a mechanism to help gauge progress, especially there is meaningful participation by those most impacted as the "strength of a community indicators measuring system is directly related to the involvement of citizens" (Sirgy et al. 2011: v).

Community indicators are unique in that when used as a system, they can support decision-making to foster improved community well-being. More communities, and even entire countries are seeking ways to improve well-being. Numerous projects at all levels exist to encourage better outcomes; many of these incorporate indicator systems as a tool for fostering deeper understanding of the connection between community development and community well-being. Examples abound—the City of Santa Monica, California (already a leader in sustainable community indicator approaches) is developing The *Local Wellbeing Index*. As they describe, the intent is to "Define, Measure, Act" by identifying and assessing indicators of well-being related to social and cultural dimensions, economics, community and connections, health, education, and the environment, combining both subjective and objective indicators (City of Santa Monica 2014). In the UK, The Local Wellbeing Project is a three-year initiative "to explore how local government can practically improve the happiness and wellbeing of their citizens" (Mulgan et al. 2008: 1). This project is a collaborative effort of three local governments working with the Young Foundation (a valuable resource for community well-being research based in London) and Professor Richard Layard from the London School of Economics (an economist who is a leading researcher in happiness and public policy studies).

In other words, the practice and discipline of community development can serve a vital role in actualizing community well-being as defined by those who live it—the citizens and residents of our villages, towns, cities and countries. Application in particular is where community development and community well-being intersect. It is a rich area worthy of more exploration to identify improved policies, programs and approaches for fostering desired outcomes.

## References

Anderson, R. E. (2014). *Human suffering and quality of life*. Dordrecht: Springer.

Anglin, R. (2011). *Promoting sustainable local and community economic development*. London: CRC Press.

Burd-Sharps, S., Guyer, P. N., Lechterman, T., & Lewis, K. (2011). The American human development index: Results from Mississippi and Louisiana. In *Community quality-of-life indicators: Best cases V*. Dordrecht: Springer.

Chanan, G. (2002). *Community development foundation measure of community: A study for the active community unit and research*. London: Development and Statistics Unit of the Home Office.

City of Santa Monica. (2014). Santa Monica well-being project. http://www.smgov.net/wellbeing/. Accessed 25 Feb 2014.

Community Development Society. (2014). Principles of good practice. http://www.comm-dev.org/. Accessed 13 Jan 2014.

Cox, D., Frere, M., West, S., & Wiseman, J. (2010). Developing and using local community well-being indicators: Learning from the experiences of community indicators Victoria. *Australian Journal of Social Issues, 45*(1), 71–89.

Haworth, J., & Hart, G. (2007). *Well-being individual, community and social perspectives*. Hampshire: Palgrave.

International Association for Community Development (IACD). (2014). http://www.iacdglobal.org/about. Accessed 30 July 2014.

Mattessich, P., & Monsey, M. (2004). *Community building: What makes it work*. St. Paul, MN: Wilder Foundation.

McKnight, J., & Block, P. (2010). *The abundant community*. San Francisco: Berrett-Koehler.

Mulgan, G., Hothi, M., Brophy, M., & Bacon, N. (2008). *Neighbourliness + Empowerment = Well-being. Is there a formula for happy communities?* London: The Young Foundation.

Online Etymology Dictionary (OED). (2014). Community. http://www.etymonline.com/index.php?term=community. Accessed 13 Jan 2014.

Phillips, R. (2003). *Community indicators measuring systems*. Aldershot: Ashgate Publishing, Ltd.

Phillips, R., & Pittman, R. (2009). *Introduction to community development*. London: Routledge.

Phillips, R., & Besser, T. (2013). *Community economic development*. London: Routledge.

Rapley, M. (2003). *Quality of life research*. Los Angeles: Sage.

Schaffer, R., Deller, S., & Marcouiller, D. (2006). Rethinking community economic development. *Economic Development Quarterly, 20*(1), 59–74.

Sirgy, J., Phillips, R., & Rahtz, D. (2011). *Community quality-of-life indicators: Best cases V*. Dordrecht: Springer.

# Searching for the Meaning of Community Well-Being

**Seung Jong Lee and Yunji Kim**

## Introduction

What do people want? What does genuine progress look like? How can government make people's lives better? These are the questions that governments around the world are becoming increasingly interested in and they are focusing on happiness and well-being for potential answers. Bhutan's gross national happiness index, the Canadian Index of Well-being, France's plans for national well-being measurement, Germany's international forum on well-being,[1]" Italy's equitable and sustainable well-being project,[2] Korea's national policies centered on citizen happiness,[3] US's plans for a national happiness index, and UK's national well-being index are just a few examples.

Interest in happiness and well-being are not new in academia. Economists, psychologists, and political scientists have built up an impressive amount of research on happiness and well-being. What *is* new in the political discourse is the desire to connect happiness and well-being to the local level through the term

[1] The international forum, "What matters to people— well-being and progress," took place on 5 June 2013 in Berlin, Germany.

[2] This project is led by the National Council for Economics and Labour and the Italian National Institute of Statistics. The first BES (benessere equo sostenible) report is available in English at the following URL: http://www.misuredelbenessere.it/fileadmin/upload/Report_on_Equitable_and_Sustainable_Well-being_-_11_Mar_2013_-_Summary.pdf.

[3] The current Park Geun Hye government has chosen citizen happiness as one of four keynote goals for national policies.

S.J. Lee (✉)
Seoul National University, Seoul, Republic of Korea
e-mail: slee@snu.ac.kr

Y. Kim
Cornell University, Ithaca, NY, USA
e-mail: yk634@cornell.edu



S.J. Lee et al. (eds.), *Community Well-Being and Community Development*, SpringerBriefs in Well-Being and Quality of Life Research,
DOI 10.1007/978-3-319-12421-6_2

community well-being (CWB). That is, even though national governments are speaking of happiness and well-being, they are emphasizing the local level as the stage wherein they can be realized. For example, the UK government has strongly encouraged local governments to become a lead partner in enhancing well-being (DETR 2000),[4] the CWB Indicators Project in Australia was launched by the Local Government Association of Queensland, and many CWB measurement projects are managed by local governments.[5]

While CWB can be a useful concept for guiding recent efforts to connect policies to well-being (Kim and Lee 2013)—particularly in the context of community development policy and practice—there is still confusion about what it means. This lack of clear definition is problematic for at least two reasons. First, definitions matter a great deal in progress of scientific knowledge for how can one study something without a clear understanding of *what* he or she is trying to study? This is why any researcher begins the research process by defining his or her research topic. Any textbook begins by defining the concepts that are to be explained in a chapter and any legal documents have a long section of definitions. Without a clear definition of what we are studying, there can be little hope of expanding knowledge through collective efforts in the scientific community.

Second, the lack of a clear definition means the term will be of little use for policy decisions. Without a clear definition, policymakers may be talking about different ideas while using the same term and this can lead to a set of policies that completely counteract each other. Even after policy decisions are made, the different understandings of the word community well-being between policymakers and citizens may lead to public disenchantment and dissatisfaction. All of these possibilities come with large costs to public resources.

The confusion around definitions of CWB becomes obvious in the casual and widespread interchanging uses with other words such as happiness, life satisfaction, quality of life, and subjective well-being. While the synonymous use of these words in everyday life may be harmless, it presents a serious problem for academics and practitioners. If indeed these terms all mean the same thing, then different studies that claim to do something new lose their validity and we would all be better off using a single term for the sake of efficiency and efficacy.

This chapter addresses this issue to provide a solid grounding for exploring community well-being. Our main questioning starts with, what is CWB and how is it different from other similar terms? We tackle the first question of defining CWB in the next section by introducing a framework for reviewing previous definitions. The second question is addressed in part three where we compare and contrast CWB with happiness, quality of life, and individual well-being.

---

[4] The Community Well-being Board has been established within the Local Government Association in the UK.

[5] See Kim and Lee (2013) for examples of CWB measurements developed and utilized by local governments.

## Definitions of Community Well-Being

What is CWB? Before looking at how the term has been defined in previous literature, we turn to a more linguistic approach to introduce a framework to guide our literature review. We focus on well-being first as the interest in CWB is grounded in the broader well-being literature.

The word "well-being" was first used as early as the 16th century but has become a buzzword of the 21st. The Merriam-Webster dictionary defines the word as "the state of being happy, healthy, or prosperous." However, the current usage of well-being seems to warrant replacing the word "or" with "and" since we do not describe a person who is merely happy, merely healthy, or merely prosperous as having a level of high well-being. Rather, the word "well-being" is used to express more than that. Scholars have often borrowed Aristotle's *eudaemonia* as being the most similar concept. The Greek term *eudaemonia* consists of the words "*eu*" and "*daimon*," meaning "good" and "spirit", respectively. The entire term is most often translated as flourishing. Although some scholars have seen *eudaemonia* as a type of well-being, along with hedonia (Henderson and Knight 2012), we follow the work of other scholars that see *eudaemonia* as more closely related to the contemporary understanding of well-being, while hedonia is closer to that of happiness (Ryan et al. 2013).

Similar to well-being, community is also a word with various meanings. In fact, Hillery (1955) found ninety four definitions of community and even contradicting definitions among them. Nonetheless, the same study found that most definitions showed agreement on the following points: community refers to persons in social interaction within a geographic area who have one or more additional common ties (Hillery 1955). Fellin (2001) identified two major types of communities—geographic and functional—and noted that both types share a common characteristic of face-to-face communication, exchange, and interaction. Thus the most general definition of community seems to involve a type of social interaction among people. We interpret the word "community" in CWB as a modifier that distinguishes it from individual well-being or national well-being. That is, we use community to refer to a geographically bound group of people on a local scale who are subject to either direct or indirect interaction with each other.

This examination of CWB concepts in two parts ("community" and "well-being") led us to the following framework for organizing previous definitions of CWB. The two terms in CWB can be used as characteristics for categorization. That is, the word "community" is related to the level of analysis and "well-being" is related to the scope of analysis. We combined these two elements as spectrums so that the "level of analysis" spectrum has individual and collective at either end, and the "scope of analysis" spectrum has partial and comprehensive at either end. Combining the two spectrums creates four quadrants: (1) collective, comprehensive; (2) collective, partial; (3) individual, partial; (4) individual, comprehensive. Figure 1 is a visual representation of this framework.

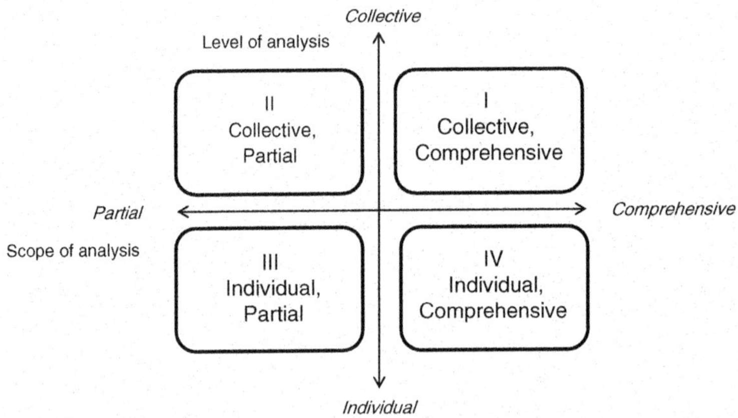


**Fig. 1** Framework for analysis of previous community well-being definitions

An important note is although we have examined the term CWB in two parts—community and well-being—this does not mean that CWB is the simple sum of community and well-being. Rather, CWB is related to community on one hand and well-being on the other. It should also be noted that the scope may include several aspects such as domain (economic, natural, physical, political, social, etc.), quantity and/or quality, objective and/or subjective, and approaches (e.g. asset, capital, emotion, resource). Thus scope indicates how many aspects are included in the analysis. In our review of previous literature, we found that most definitions of CWB have only focused on the domain aspect of the scope of analysis. However, this result does not mean that the scope of CWB connotes only domain aspects.

Using the above framework, we analyzed the following literature found through searches in SociINDEX, ASSIA (Applied Social Sciences Index and Abstracts), Sociological Abstracts, and Google Scholar. We narrowed our review to literature that specifically use the term CWB and define it. Interestingly, we found that none of these studies defined CWB as limited to a narrow scope at an individual level (category III). Instead, most of the studies that use the definition of category III were labeled as quality of life studies and focused on the socioeconomic domain at the individual level.

The most limited definitions of CWB appear in quadrant IV. For example, Prilleltensky and Prilleltensky (2006) defined CWB as consisting of physical, geographic, cultural, economic, political, and psychosocial environments where community members have their needs met. Although this definition acknowledges diverse factors of CWB it still focuses on the needs of individual members, locating it in category IV.

On the other hand, McHardy and O'Sullivan (2004) and Allensworth and Rochin (1996) focus on the socioeconomic needs of community. However, these definitions are also narrow in the sense that CWB is limited to socioeconomic conditions only. These definitions can be placed in category II.

A more holistic approach includes a variety of community factors, such as physical, psychological, political, social, cultural, and environmental. These are the definitions in category I. The City of Calgary (2010) defines CWB as incorporating economic, social, and physical well-being. Cox et al. (2010) takes an even more comprehensive approach by including economic, social, environmental, cultural and governance goals and priorities. Chanan (2002) also defines CWB as a multifaceted term while acknowledging the innate abstractness of the concept:

> [community] [is] a number of people who have some degree of common identity or concerns often related to a particular locality or conditions … a community is not a thing. It is a number of people who have repeated dealing with each other. When community is identifiable with a locality, CWB/the quality of community life is intimately connected with: how well that locality is functioning; how well that locality is governed; how the services in that locality are operating; and how safe, pleasant and rewarding it feels to live in that locality.

The Rural Assistance Information Network (2004) in Australia states that CWB is a concept that refers to an optimal quality of health community life, which is the ultimate goal of all the various processes and strategies that endeavor to meet the needs of people living together in communities. It encapsulates the ideals of people living together harmoniously in vibrant and sustainable communities, where community dynamics are clearly underpinned by 'social justice' considerations.

The most popular use of CWB combines quadrant I and IV, looking at both individual and collective levels. For example, Cuthill (2002) defines CWB as perceptions of life in a community and explains that "description or measurement of these perceptions takes into consideration both qualitative and/or quantitative data of natural, physical, financial, social and human capital which influence both citizen's and community well-being." Hay et al. (1996) focus more on the desires of citizens and define CWB as the fulfillment of the aspirations of different individuals and groups in society. Brasher and Wiseman (2008) and Kusel and Fortmann (1991) also identify the various conditions identified by individuals and the community as community well-being. Ribova (2000) sees the concept as a framework for community assessment that recognizes the psychological, cultural and social requirements of people, and their communities. The Nuclear Waste Management Organization (NWMO 2009) of Canada also mentions diverse components at the individual and community level:

> [the term CWB] includes a combination of abstract ideas and human actions…Concepts of community well-being may reflect the interests of individuals within a community and they may also reflect the interests of the collective of community interests. Concepts of well-being may encompass social, economic, spiritual and cultural factors, as well as individual health and security.

These previous interpretations and their respective category are summarized in Table 1.

Important points emerge from reviewing these previous definitions of CWB. First, although there are various definitions of CWB, they all refer to needs, desires, aspirations, or goals. Therefore the concept refers to what is necessary in people's

**Table 1** Definitions of community well-being

| Author | Definition | Category |
|---|---|---|
| Prilleltensky and Prilleltensky (2006) | Physical, geographic, cultural, economic, political, and psychosocial environments where community members have their needs met | IV |
| Allensworth and Rochin (1996) | Socioeconomic well-being of communities | II |
| McHardy and O'Sullivan (2004) | Socioeconomic conditions of communities | II |
| Chanan (2002) | How well that locality is functioning; how well that locality is governed; how the services in that locality are operating; how safe, pleasant and rewarding it feels to live in that locality | I |
| City of Calgary (2010) | Incorporating economic, social, and physical well-being | I |
| Cox et al. (2010) | Economic, social, environmental, cultural and governance goals and priorities identified as important by a community, population group or society | I |
| Rural Assistance Information Network (2004) | Optimal quality of healthy community life … that encapsulates the ideals of people living together harmoniously in vibrant and sustainable communities, where community dynamics are clearly underpinned by 'social justice' considerations | I |
| Cuthill (2002) | Perceptions of life in a community. Description or measurement of these perceptions takes into consideration both qualitative and/or quantitative data of natural, physical, financial, social and human capital which influence both citizen's and community well-being | IV + I* |
| Hay et al. (1996) | The fulfillment of aspirations of different individuals and groups in society | IV + I |
| Kusel and Fortmann (1991) | Economic, social, cultural and political components of a community in maintaining itself and fulfilling the various needs of its local residents | IV + I |
| Nuclear Waste Management Organization (NWMO) of Canada (2009) | Combination of abstract ideas and human actions…Concepts of community well-being may reflect the interests of individuals within a community and they may also reflect the interests of the collective of community interests. Concepts of well-being may encompass social, economic, spiritual and cultural factors, as well as individual health and security | IV + I |

(continued)

**Table 1** (continued)

| Author | Definition | Category |
|---|---|---|
| Ribova (2000) | Framework for community assessment that recognizes the psychological, cultural, and social requirements of people and their communities | IV + I |
| Brasher and Wiseman (2008) | Combination of social, economic, environmental, cultural, and political conditions identified by individuals and their communities as essential for them to flourish and fulfill their potential | IV + I |

*IV + I: Combination of both "individual, comprehensive" and "collective, comprehensive"

lives as well as what is desirable. The inclusion of aspirations indicates that the factors of CWB may change as a community's preferences change.

Second, CWB is a more comprehensive concept that includes environmental, physical, and political domains in comparison to the previously limited concepts of quality of life that tend to focus simply on the socioeconomic domain or even just economic growth. This is in line with the origin of social indicators that attempted to correct the imbalance of economic indicators. However, there are still definitions of CWB that are partial as they focus only on the individual level or socioeconomic factors (category II or IV).

Lastly, there is a tendency to conflate CWB with individual well-being as can be seen in the literature that combine quadrant IV and I. We see this to be problematic for the following two reasons. First, the use of CWB to refer to two concepts that can conflict with each other may lead to theoretical conflicts. For example, Lindsay (1995) points out that individual well-being and CWB can be in conflict with respect to automobile use and the natural environment. The use of automobiles may enhance individual well-being but the consequences of too many people using automobiles instead of public transportation results in greater carbon gas emission, greater consumption of petroleum, and thus lower CWB.

Second, the interchanging use of two concepts at different levels gives little direction for practical policy decisions. We see CWB as the most appropriate concept to guide local governments because it is both impractical and unrealistic for governments to focus on directly enhancing individual well-being. Just as individual well-being and CWB can conflict with each other, there can be multiple points of conflict among individual well-being in a society. Moreover, some scholars of happiness research argue happiness is an unalterable trait that have strong links to genetic makeup (Ebstein et al. 1996; Hamer 1996; Lykken and Tellegen 1996; Tellegen et al. 1988; Lieberman 1970). Thus the more practical course of action for governments is to enhance CWB that will hopefully enhance individual well-being. In other words, we can hypothesize a causal relationship between CWB and individual well-being. However, this relationship can only be conceptualized when individual well-being and CWB are identified as distinct concepts. Without this distinction, there is confusion about the direction of this

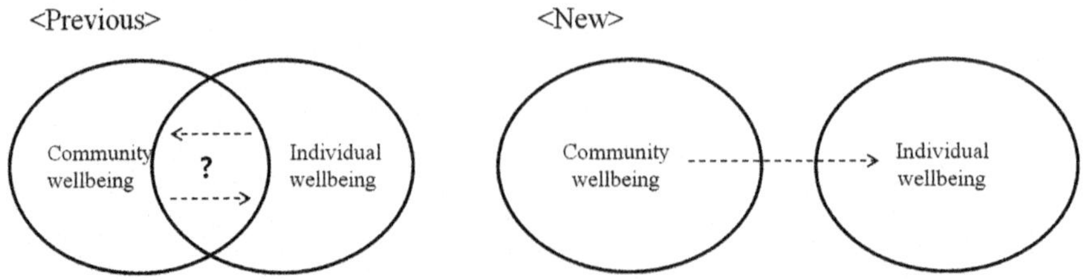


**Fig. 2** Hypothetical relationship between individual well-being and community well-being

causal relationship since the cause and effect are seen as overlapping. Figure 1 is a visual explanation of this relationship (Fig. 2).

Based on these issues, we argue that although there are different definitions of CWB, the core idea of this concept is best represented by category I as a collective concept. By collective concept we mean that CWB is more than the sum of individual well-being. To be clear, we are not arguing that individual well-being and CWB have no relationship. In fact, they are closely related and we elaborate on the relationship between the two in the next section. Another key point is that this does not mean the measurement of CWB should not be solicited from individuals. After all, individuals are what make up a community and as Prilleltensky and Prilleltensky (2006) argue CWB can be observed on an individual level, organization level, and community level.

## Community Well-Being, Happiness, Quality of Life, Community Development, and Well-Being[6]: What's the Difference?

The previous section surveyed existing definitions of CWB. This section discusses the meaning of CWB by comparing and contrasting it with other related terms, such as happiness, quality of life, and individual well-being. As mentioned before, we see the interchanging use of these terms as cause for concern. However, the interchanging use is certainly understandable given the history of how these concepts emerged. The following is a brief description of that history.

---

[6] The use of "well-being" alone most often refers to individual well-being, while other types or levels of well-being are modified with other words such as collective well-being, social well-being, and community well-being. We follow this convention of using well-being to refer to individual well-being.

The term CWB emerged in the midst of a larger movement called the social indicators movement. Large scale government involvement in citizen welfare began after World War II when the devastation of the World Wars pushed many national governments to initiate projects and policies focused on economic development. The use of gross domestic product became popular as *the* measure of progress. However, the shortcomings of economic indicators as the sole measure of progress soon sparked the social indicators movement in the 1960s. The movement was closely connected to the emergence of quality of life studies that emphasized social costs and quality, instead of mere quantity. By the 1980s, the social indicators movement briefly lost momentum as it was shadowed once again by enthusiasm for economic growth, but was revived in the 1990s as concerns of social justice, equity, and freedom increased.

Another strand of studies, called the happiness studies, began to take form in the 1970s with the well-known Easterlin Paradox[7] (Easterlin 1974). Mostly economists and psychologists have searched for factors that influence, or do not influence, happiness with subjective well-being or life satisfaction data. These data are collected from surveys that ask respondents to rate their happiness or life satisfaction levels. They have most frequently used the term well-being interchangeably with happiness, life satisfaction, or subjective well-being and we also treat these three terms as being synonymous.

All of these terms have been used interchangeably and certainly share similar goals of "making society better." But what are the differences? If there is no distinction among these terms than academics would be better off settling on a single term. The following discussion aims to show that there are indeed differences among these terms. The purpose is to arrive at a clearer understanding of CWB by distinguishing it from other terms that have previously been considered synonymous.

We begin with the most dissimilar pair and proceed to the more similar pair in the following order: CWB and happiness; well-being and quality of life; and finally CWB and well-being. For comparison, we focus on the various aspects of the scope of analysis such as how these concepts are measured, how they are conceptualized, and how they offer policy implications.

The first comparison is between CWB and happiness. In the previous section of this chapter we have seen that CWB refers to the fulfillment of the needs and desires of a community. How is happiness defined? According to the Merriam-Webster dictionary happiness is "a state of contentment or a pleasurable or satisfying experience." The key words used are content, pleasurable, and satisfying. In short, happiness is a concept that is connected to emotions and thus is heavily researched in the psychology sector. In terms of domain, happiness has focused mostly on the

[7] This Paradox refers to the phenomenon in which increasing levels of wealth were not connected to increasing levels of happiness.

psychological domain while CWB encompasses diverse domains of cultural, economic, environmental, psychological, physical, political, and social. Economists have also studied happiness from a utilitarian perspective and thus the concept of happiness is focused on the quantitative level of positive emotions, rather than quality. In contrast, CWB is interested in both quantity and quality of its components.

Most happiness studies have treated happiness as synonymous with life satisfaction and subjective well-being to emphasize the emotional, personal characteristics of this concept (Veenhoven 2012). Thus measurements of happiness are more concerned with subjective[8] evaluations on an individual level, while CWB encompasses both objective and subjective evaluations at the collective level.

As the definition indicates by the use of the word "state," happiness is a more static concept. In contrast, CWB contains the word "being," which indicates a more dynamic concept that is in motion and thus acknowledges and emphasizes the process towards an end goal. Conceptually, happiness takes an emotional approach while CWB takes an asset approach (e.g. practical efforts to measure CWB often include an asset mapping step).

For policy guidance, happiness aims to induce more positive emotions while CWB aims for more production, accumulation of assets and ultimately flourishing by realizing these community assets. The different goals that happiness and CWB suggest are reflections of their value judgment. That is, happiness as a concept is value-neutral and simply accepts that more positive emotion is desirable regardless of how this emotion is achieved. In contrast, CWB is a value-driven concept that makes implications of whether a certain factor is good or bad for CWB.

Next, we compare well-being and quality of life. Everyday usage and even some scholarly works seem to suggest that these two concepts are in fact the same thing (Galloway 2006). We argue here that these terms can be differentiated and warrant different names. Quality of life was popularized as a term that could correct the bias towards quantity, especially in terms of economic wealth. As such, the concept has emphasized measuring the quality of social and economic factors in an objective manner. In comparison, well-being focuses on both quantity and quality of its factors and incorporates both objective and subjective measurements. These characteristics are in line with the view that well-being is a more comprehensive concept than quality of life. One similarity for well-being and quality of life is the focus on individuals as the level of analysis. For example, Scott (2012) points out that quality of life has been promoted by neoliberal discourse as "belonging to autonomous individuals which could be enhanced in the market place."

[8] The comparison of quantity and quality should not be confused with that of objective and subjective. The former refers to the characteristic of an object while the latter concerns the method of evaluation. For example, a community's medical service can be high in quantity (e.g. number of hospitals) but low in quality (e.g. patient satisfaction of doctor visits). The quantity of this aspect can be evaluated both objectively (e.g. comparison to the average number of hospitals in communities of comparable size) and subjectively (e.g. resident evaluation of the number of hospitals).

Similar to CWB, well-being is also a dynamic concept while quality of life is a more static concept. Quality of life is not focused on process but rather the current level of quality of life is adequate. Well-being also takes an asset approach as CWB, while quality of life takes a capital approach. We borrow the terms "asset" and "capital" from finance to illustrate the different emphasis on the "how to" enhance quality of life and well-being. Capital is the more narrow term that refers to a financial asset, such as cash that is required to produce goods[9] while asset is a broader term that refers to things that have value in and of itself. Quality of life focuses on measuring the end-level of capital while well-being focuses on a person's assets or potential that has value but perhaps has not been capitalized yet.

The different approaches also connect to the different end goals that quality of life and well-being suggest. Since quality of life is focused on objective measures of the qualitative aspects of life, it tends to focus on fulfilling any deficiencies in these aspects. In contrast, well-being focuses on the production, accumulation, and flourishing of individual potential. Moreover, since well-being is similar to Aristotle's *eudaemonia* the end goal of flourishing is value-driven while quality of life is more value-neutral.

As previously mentioned, well-being and CWB share several basic characteristics. In terms of measurement, they both focus on quantity and quality of factors, and include both objective and subjective evaluations. In addition, they are both dynamic concepts that take an asset approach with implications for value-driven flourishing as the end goal. The distinguishing characteristic between these two concepts is the level of analysis. In other words, well-being is focused on the individual while CWB is a collective term. While this may be a simple difference, the mixing of these two concepts can be dangerous both theoretically and practically as we have examined in the previous section.

CWB is also deeply connected to community development as residents and local government alike are very concerned with this dimension. As discussed in chapter, "Exploring the Intersection of Community Well-Being and Community Development", community development is defined as both a process and an outcome for across a range of considerations (physical, social, cultural, political, environmental, etc.). Community development builds on the concept of assets or capital in an area across all realms. Within community development, we can see the direct relationship to community well-being, as both a goal/outcome and processes of progressing towards desired states of well-being.

Table 2 summarizes the characteristics of happiness, quality of life, well-being, community development, and CWB that have been discussed thus far.

---

[9] As defined on www.investopedia.com.

**Table 2** Comparison of community well-being and related concepts

| | Community well-being | Happiness | Quality of life | (Individual) Well-being | Community development |
|---|---|---|---|---|---|
| Individual/ Collective | Collective | Individual | Individual | Individual | Collective |
| Domain | Cultural, economic, environmental, social, physical, political | Psychological | Economic, social | Economic, social, physical, psychological | Cultural, economic, environmental, social, physical, political |
| Quantity/ Quality | Both | Quantity | Quality | Both | Both |
| Objective/ Subjective | Both | Subjective | Objective | Both | Both |
| Static/ Dynamic | Dynamic | Static | Static | Dynamic | Dynamic |
| Approach | Asset | Emotion | Capital | Asset | Asset |
| Goal | Production; accumulation; flourishing | Induce positive emotion | Fulfillment of deficiency | Production; accumulation; flourishing | Production; accumulation; flourishing; fulfillment of deficiency |
| Value judgment | Value-driven | Value-neutral | Value-neutral | Value-driven | Either |

## Conclusion

CWB has become a buzzword in the policy world and researchers are trying to keep up with this popular interest. While there has been a considerable amount of research on the topic, many of these studies have been carried out without clarifying the meaning of CWB. Instead, there has been a casual interchanging use of CWB with happiness, quality of life, and individual well-being. We see this as a serious impediment to advancing the academic dialogue and use of CWB for policy guidance. The confusion around the concept limits hypothesizing and testing relationships between CWB and other terms as well as creating inefficiency in research and communication. As such, this chapter has sought to clarify the meaning of CWB by first surveying how the term has been defined in the past and by distinguishing it from happiness, quality of life, and individual well-being.

A survey of previous definitions of CWB showed that while there are several definitions, they emphasize needs and desires in common. In addition, most definitions of CWB were comprehensive, encompassing several domains such as social, economic, cultural, environmental, and political. We also found that many previous definitions of CWB have the tendency to conflate CWB with individual well-being and pointed out that this is problematic as there can be instances of individual well-being and CWB conflicting and the confusion that arises in conceptualizing the relationship between the two.

A comparison of CWB with happiness, quality of life, community development and well-being also shows that CWB is a more comprehensive concept. For example, CWB focuses on both quantity and quality of factors and utilizes both objective and subjective measurements. In contrast, happiness focuses solely on quantity and subjective measurements while quality of life focuses on quality and objective measurements. We also pointed out that CWB is a more dynamic concept that takes an asset approach and points to the value-driven goal of flourishing, much akin to community development.

Based on a conceptual and theoretical discussion of CWB, we proposed a definition of CWB that has a comprehensive scope at the collective level. However, there is still more work to be done because different communities can define CWB differently (NWMO 2009). For instance, in the previous section, we identified some aspects of the scope of CWB such various domains, quantity and/or quality, objective and/or subjective, and asset or capital approaches. Future studies of CWB may identify and include additional aspects, leading to more refined definitions of CWB.

These works on defining CWB should be accompanied by empirical research for the concept to be practically useful. For example, our proposed relationship between CWB and individual well-being can be tested in different contexts, and the influencing factors of CWB can be identified as well as the relationships among these factors. The combination of these theoretical and empirical works are both necessary to build a more sophisticated model of CWB.

## References


Allensworth, E. M., & Rochin, R. I. (1996). *White exodus, Latino repopulation, and community well-being: Trends in California's rural communities*. JSRI Research Report No. 13.

Brasher, K., & Wiseman, J. (2008). *CWB in an unwell world: Trends, challenges, and opportunities*. Policy Signpost 1. McCaughey Centre.

Chanan, G. (2002). *Community development foundation measure of community: A study for the active community unit and research*. London: Development and Statistics Unit of the Home Office.

City of Calgary. (2010). *Indices of community well-being, 2006 for Calgary neighbourhoods*. Calgary, Alberta: The City of Calgary.

Cox, D., Frere, M., West, S., & Wiseman, J. (2010). Developing and using local community well-being indicators: Learning from the experiences of community indicators Victoria. *Australian Journal of Social Issues, 45*(1), 71–89.

Cuthill, M. (2002). Coolangatta: A portrait of community well-being. *Urban Policy and Research, 20*(2), 187–203.

DETR. (2000). *Local quality of life counts: A handbook for menu of local indicators of sustainable development*. London: Department of the Environment, Transport and the Regions.

Easterlin, R. (1974). Does economic growth improve the human lot? In P. A. David & M. W. Reder (Eds.) *Nations households and economic growth: Essays in honor of Moses Abramovitz*. New York: Academic Press.

Ebstein, R. P., Novick, O., Umansky, R., Priel, B., Osher, Y., Blaine, D., et al. (1996). Dopamine D4 receptor (D4DR) exon III polymorphism associated with the human personality trait of novelty seeking. *Nature Genetics, 12*(1), 78–80.

Fellin, P. (2001). *The community and the social worker* (3rd ed.). Itasca, IL: F.E.Peacock.

Galloway, S. (2006). *Quality of life and well-being: Measuring the benefits of culture and sport: A literature review*. Edinburgh: Scottish Executive Social Research.

Hamer, D. H. (1996). The heritability of happiness. *Nature Genetics, 14*(2), 125–126.

Hay, D. I. (1996). *Keep it grounded and keep it simple: Measuring community well-being*. Vancouver: Social Planning and Research Council of British Columbia.

Henderson, L. W., & Knight, T. (2012). Integrating the hedonic and eudaimonic perspectives to more comprehensively understand wellbeing and pathways to wellbeing. *International Journal of Wellbeing, 2*(3), 196–221.

Hillery, G. A. (1955). Definitions of community: Areas of agreement. *Rural Sociology, 20*, 111–123.

Kim, Y., & Lee, S. J. (2013). The development and application of a community wellbeing index in Korean metropolitan cities. *Social Indicators Research* [Online first publication], 1–26. doi:10.1007/s11205-013-0527-0.

Kusel, J., & Fortmann, L. (1991). *Well-being in forest-dependent communities* (vol. 1). Berkeley, CA: Forest and Rangeland Resources Assessment Program and California Department of Forestry and Fire Protection.

Lieberman, L. R. (1970). Life satisfaction in the young and the old. *Psychological Reports, 27*(1), 75–79.

Lindsay, H. (1995). Balancing community needs against individual desires. *Journal of Land Use and Environmental Law, 10*(2), 371–403.

Lykken, D., & Tellegen, A. (1996). Happiness is a stochastic phenomenon. *Psychological Science, 7*(3), 186–189.

McHardy, M., & O'Sullivan, E. (2004). *Five nations community well-being in Canada: The community well-being index (CWB), 2001*. Indian and Northern Affairs Canada: Strategic Research and Analysis Directorate.

Nuclear Waste Management Organization (NWMO). (2009). *Applying community well-being: Lessons and experience of canadian practitioners*. NMWO Report SR-2009-02 2009.

Prilleltensky, I., & Prilleltensky, O. (2006). *Promoting well-being: Linking personal, organizational, and community change*. Hoboken, NJ: Wiley.
Ribova, L. (2000). *Individual and community well-being*. The Arctic.
Rural Assistance Information Network. (2004). Retrieved from http://www.rain.net.au/community_well-being.htm.
Ryan, R. M., Huta, V., & Deci, E. L. (2013). Living well: A self-determination theory perspective on eudaimonia. In A. Delle Fave (Ed.), *The exploration of happiness*. Dordrecht: Springer.
Scott, K. (2012). *Measuring wellbeing: Towards sustainability?*. London: Routledge
Tellegen, A., Lykken, D. T., Bouchard, T. J., Wilcox, K. J., Segal, N. L., & Rich, S. (1988). Personality similarity in twins reared apart and together. *Journal of Personality and Social Psychology, 54*(6), 1031.
Veenhoven, R. (2012). Happiness, also known as "life satisfaction" and "subjective well-being". In K. Land, A. C. Michalos, & M. J. Sirgy (Eds.), *Handbook of social indicators and quality of life research*. Dordrecht: Springer.

# Building Community Well-Being Across Sectors with "For Benefit" Community Business

**Rhonda Phillips**

## Introduction

In the previous chapter of this volume, a community well-being framework is presented, illustrating the overarching reach of community well-being's foundational aspects. This chapter provides an illustration of the connection to two of the major foundational aspects of the framework—sustainability and community development. Community business is reflective of both these areas. This chapter provides further examples of using tools and techniques, as reflected in community indicators, to aid in progress towards promotion of overall community well-being and development.

Another purpose is to provide an illustration of an alternative way of thinking about these concerns related to economic and social well-being, from the vantage point of community well-being. It is proposed that benefits accrue in an environment where social, natural and cultural dimensions of community are important and recognized by the private sector as well as the civic sector (public and non-governmental organizations) and reflected in community-based business practices. For example, when socially responsible or socially focused enterprises locate in or generate from within an area, it serves to attract others similarly oriented to grow their organizations within the community. These socially focused enterprises can be private for profit businesses, nonprofit/nongovernmental organizations, or public sector agencies. This broad definition—for benefit community business—implies those entities conveying positive externalities for community economic development processes and outcomes when these socially minded enterprises are participating partners. The following sections provide insight into the issue of locally focused economies, propose ideas for a framework to consider, and presents the case of Burlington, Vermont in the United States as a progressive economy centering on locally focused and socially responsive businesses and enterprises. It

R. Phillips (✉)
Purdue University, West Lafayette, IN, USA
e-mail: rphillips@purdue.edu



S.J. Lee et al. (eds.), *Community Well-Being and Community Development*,
SpringerBriefs in Well-Being and Quality of Life Research,
DOI 10.1007/978-3-319-12421-6_3

concludes with policy implications for others to consider who may be interested in fostering community well-being via locally focused community business and enterprise.

## Why Local?

With a focus on locally owned businesses, socially responsible and vested organizations, and a preference for governance versus government, some areas are embracing the movement towards a "local economy." This is seen in such activities as the Local First efforts in the US where local products, services and enterprises are promoted actively via buy local campaigns and similar programs. It is about more than buying local, however. This shift in perspective and orientation is concerned with creating a community enterprise and business culture that survives short-term economic and political changes and endures through time—in other words, it becomes more sustainable and helps foster community well-being. A locally focused economy brings together many of these dimensions including socially responsible enterprises and culturally competent businesses and organizations to aid in development of a long term, enduring economy (Phillips et al. 2013).

Co-founder of the Business Alliance for Local Living Economies (BALLE) Michael Schuman explains in his book *The Small-Mart Revolution* that local ownership is an essential condition for community prosperity for at least the following five reasons:

1. Locally-focused businesses are long-term wealth generators
   Many entrepreneurs are in a particular community because they love living there and this makes them less likely to leave. The longevity of some of these businesses can span several generations.
2. Fewer destructive exits from a community's economy
   Massive upheavals can occur in a local economy when a large employer exits, creating a "death spiral" where a sudden exit is followed by high levels of unemployment, shrinking property values, lower tax collections, deep cuts in schools, police and other services. Economies comprised of locally owned businesses are far less likely to experience this dramatic decline.
3. Higher Labor and Environmental Standards
   Local quality of life is better protected in communities comprised mostly of locally owned businesses, via shaping of its laws, regulations and business incentives. Locally owned businesses do not typically threaten to leave town, and can set reasonable labor and environmental standards with confidence. Further, business incentives can be tailored to the needs of the majority of locally owned businesses, rather than giving large subsidies to nonlocal businesses. Typically, locally-focused businesses are more responsive and more vested in the community.

4. Better chances of success
   Locally owned businesses are not as susceptible as large companies to temptation to move when costs rise—witness the lure of off-shore manufacturing locations for many industries throughout the US. Moving or relocating to another community or country usually just is not an option for many locally owned businesses.
5. Higher economic multipliers
   Studies show that the impact of a dollar spent at local businesses has a far greater impact than money spent at chains, big box, or formula retail. Local businesses yield two to four times the multiplier benefit (reiterative spending in an economy) as compared to nonlocal businesses. Local businesses have higher multipliers because they spend more locally. In other words, local management uses local services, advertise locally, and enjoy profits locally. To illustrate, there is only one franchise in the National Football League owned by a community controlled nonprofit with shareholder members (primarily residents of Wisconsin). While other franchises can and do leave their host communities, the Green Bay Packers are a critical source of wealth and economic multipliers for Green Bay and will be around for many years (Schuman 2007 as cited in Phillips et al. 2013, 29–31).

There is another reason as well. Strong local economic ownership can improve an area's prosperity because these type enterprises support the transition to a more sustainable economy that is simultaneously locally and globally oriented. If local government policy supports focusing on producing more needed goods and services otherwise imported at a higher total cost such as food sources, then the economy will strengthen showing signs of being more responsive and resilient. Those local and regional governments able to respond to these opportunities are emerging as leaders with policies and programs providing things people need locally rather than importing the majority of these products and services. This in turn leads to greater community well-being.

Many of these socially oriented enterprises and socially responsive businesses start off small, generated from within a local economy, or attracted there because of similarly focused enterprises already in the area. These vested organizations tend to be tied more closely with and committed to their host community. This in turn furthers the connections of trust, reciprocity, and support from within the community—these are the basic foundations of good community development practice. The connections between locally focused economies and socially responsible businesses are strong. Locally generated and owned businesses often will reinvest in their communities beyond the range of their own business arena. Social, environmental, educational and other dimensions of community are important and are reflected in the businesses' and enterprises' activities and initiatives (Phillips et al. 2013). This often gives rise to a long and enduring history of social responsibility in a community with these type enterprises.

This approach is also reflective of good practice in community development as it is highly focused on the issue of building social capacity. The more social capacity a community has, the more likely it can adapt to and work around deficiencies in

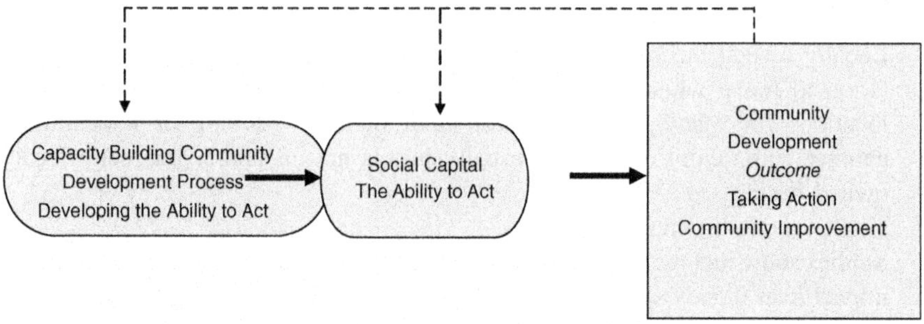


**Fig. 1** The capacity building—community development link. *Source* Phillips and Pittman (2009, p. 7)

the other types of community capital in the areas of the economy and environment (Phillips and Pittman 2009). How is social capital created or encouraged? The process of community development is social capital/capacity building which leads to social capital, which in turn leads to the outcome of community development. The progression of this is noted in the following chart. Solid lines show the primary flow of connection while feedback loops are represented by the dotted lines. As shown in the third box, progress in the outcome of community development (taking positive action resulting in improvements in the community) contributes to capacity building (the process of community development) and social capital (Fig. 1).

It takes many types of organizations—public, private and nonprofit—to help a community thrive and build social capital. In the recent past, there has been a "blurring of lines" across these organizational types and sectors as societal needs and desires change, and new approaches are emerging. The active transformation of what is known as the Third Sector (predominately nonprofit or non-governmental organizations as well as others socially focused) is providing more options to accomplish improvements in community well-being. Heerad Sabeti describes the advent of a new generation of organizations at the intersection of public, private and social sectors as follows, in what some are referring to as the new "Fourth Sector":

a. The convergence of organizations toward a new landscape—a critical mass of organizations within the three sectors have been evolving, or converging, toward a fundamentally new organizational landscape integrating social purposes with business methods;
b. The emergence of hybrid organizations—pioneering organizations have emerged with new models addressing societal challenges blending attributes and strategies from all sectors. They are creating hybrid organizations transcending the usual sectoral boundaries and resisting easy classification within the three traditional sectors (Sabeti 2009).

Sectors have begun to merge even more in the recent past, due to motives and desires of the organizations involved. The following illustration provides a look at the terms centering on issues of social responsibility and social enterprise as well as the areas of interface between the sectors. What is encouraging about these newer activities, regardless of the name prescribed (Third or Fourth Sector) are the implications they hold for enhancing and fostering community well-being. Bringing together resources and energy across sectors holds potential for a larger collective impact on communities. Additionally, it highlights the contributions and potentials of the social/civic sectors to address issues that influence communities and their well-being on a variety of fronts.

## Crafting a Framework

There are numerous types of key data to consider in approaches for encouraging and supporting community businesses and social enterprises. These indicators can be adapted to various community environments and conditions, and can help leaders and residents foster strategic approaches for promoting community well-being. Each community will have its own values and subsequent goals reflecting these. However, the following ideas are presented as potential indicators in a community well-being framework focusing on fostering socially responsible businesses and enterprises. It combines the key concepts of community development, as well as social enterprise and supportive civic policy. A framework such as this can be used to help discern areas where additional attention is needed, as well as gauge existing strengths in a community desiring to encourage locally focused community economic development.

*Economic Indicators* Economic diversity (which sectors account for the majority of the economic activity); quality of jobs (in relation to cost of living and skills of the area's residents); jobs/housing balance (number of jobs in relation to housing units and cost of housing); quality of jobs (wages in relation to livable wages and cost of living in community); types of innovative/alternative enterprises (green/sustainable or particular sectors that are regarded as socially impactful).
*Socially Responsive Business Indicators* Business reinvestment in community; percent businesses majority locally owned versus absentee ownership (including formula retail stores versus locally owned, for example); types of giving and community support; ownership structures; level and type of activities in sustainability/socially responsiveness (corporate social responsibility and related).
*Social Enterprise Indicators* Numbers and types of nonprofit organizations; numbers and types of alternative structure organizations (for benefit corporations, cooperatives, worker-owned organizations, etc.); longevity and generation of social enterprise organizations (to gauge local environment of support).
*Community Development Indicators* Level and type of participation by organizations in civic society; interface with residents (for example, shared facilities, public space, activities and opportunities for interaction); strength of response to issues facing society (level and pervasiveness of major issues in the community whether environmental, social, economic or a combination); availability of products and services generated in the local area (types including green/sustainable such as local food, arts and cultural products and services).

*Supporting Policy Indicators* Policies in place supporting locally focused enterprises and businesses; level and type of support for infrastructure, programs and activities; responses to issues impacting the ability to foster a locally focused economy.

Combining these indicators with an assessment of the situation within the community on factors such as strengths of the local economy will provide a more comprehensive picture. Factors can be gauged such as: what is the community noted for including any specialties (local foods or arts and cultural services or green/sustainable enterprises, for example); what types of enterprises are currently located there; and what types of activities are attracted to the area and why; is there a culture of participation by both individuals and organizations in the community; and what is the nature of approaches to solving major social issues (predominately public sector or a partnership approach, for example). Of course, calibrating to a particular local area requires reflecting values and preferences of residents will be essential. Having engaged and meaningful participation and reflection of these values and preferences will help selection and use of indicators most appropriate for a community.

## The Case: Burlington, Vermont[1]

Burlington is located in the Northeastern region of the United States, just south of the Canadian border. It is noted as a progressive environment and has attracted and generated alternative and sustainable business and civic sector enterprises over the last 50 years. Often cited as a leader in sustainable community economic development, Burlington with its 42,000 residents is Vermont's largest city. The metropolitan area is home to over 200,000 representing one-third of the entire population of the state.

Attention on locally focused economies has always been strong in Burlington and indeed, the entire state of Vermont. The Local First movement emerging across the US has been a mainstay for many years in the area. Local First Vermont is the nonprofit organization located in the Burlington area focused on preserving the character and prosperity of Vermont's economy, community networks and natural landscape. Their mission is to preserve and enhance the economic, human and natural vitality of Vermont communities by promoting the importance of purchasing from locally-owned, independent businesses (Local First Vermont 2012). They sponsor a variety of education programs as well as enhancing marketing efforts, with Local First Vermont decals in merchants' storefronts or by offering special coupon books to spur interest in locally focused businesses. As mentioned, Local First initiatives across the US are increasing rapidly, and often include

[1] This section is excerpted and adapted from the forthcoming book, Progressive Capitalism, by R. Phillips, B. Seifer and E. Antzcak.

promotional efforts for locally focused businesses and enterprises with special marketing, such as a Buy Local coupon book.

## *Growing Socially Responsive Businesses*

Vermont has long been an innovator in socially responsible businesses. They were the first state in the US to form a nonprofit association of socially responsible business owners, the Vermont Businesses for Social Responsibility (VBSR). This organizations strives to strengthen the social and environmental infrastructure in the state by fostering a climate to support sustainable business growth. Vermont Businesses for Social Responsibility has members representing a variety of sectors across the state. All together, 1,200 members employ more than 13 % of Vermont's workforce and generate more than US$4 billion in revenue annually; 60 % of members have been in business more than 10 years (VBSR 2012).

VBSR promotes business ethics and business models that value multiple bottom lines that correspond with sustainability precepts: economic, social, and environmental dimensions. Their mission is undertaken by the following activities:

1. *Education* Empowering members to solve environmental, social and economic problems by providing concrete resources and information to help improve members' business practices.
2. *Public Influence* Initiating tangible change in public policy combating exploitation and promoting sustainability by working to capture the inherent power of business to represent socially responsible ideals to legislative bodies, news media, and the general public.
3. *Workplace Quality* Enabling every worker to live and work with respect and dignity by creating work environments and economic climates providing fair income in safe work settings, and allowing each employee to contribute to a high quality product or service (VBSR 2012 as cited in Phillips et al. 2013, 38).

The ability of VBSR to influence public policy in the state is strong. Using an approach to analyze policy impacts, their lobbying efforts focus on state level policy. Several of their accomplishments include lobbying for the sustainable jobs fun, supportive family leave legislation (which should be noted that this served as a national model adopted by the US. Congress a year later), and farm-to-plate funding for encouraging farm initiatives. Over 20–25 policies are annotated on their website showing the breadth of their policy initiatives. It is their approach to be never against anything and always for something, from a policy perspective (39). They provide an annual Legislative Scorecard, for members and others to see how legislators voted on each year's VBSR's business agenda.

Phillips et al. (2013, 40) describe other benefits of engaging in socially responsible business practices as,

> "it tends to help set a positive tone for an area. Benefits accrue in an environment where social, natural and cultural dimensions of community are important and recognized by the private sector. When socially focused or socially responsible businesses locate or generate from within an area, it serves to attract others of like mind to grow their businesses in the community."

Local business and industry need support, encouragement, incentives and control. As John Abrams, author of *The Companies We Keep*, and founder of a community and social focused business in New England, explains:

> "Fortunately, despite whatever encroachments have been made by dispassionate big business (each new Wal-Mart ultimately puts an astonishing number of local businesses under), we still have our local economies. We don't need to take them back from global corporations; we already have them, in whatever condition they're in—good, fair, or poor. We can move them forward from here. And if we can keep local institutions and businesses alive… we will keep the continuity of generations alive and maintain the richness of our communities" (Abrams 2008, 15).

Abrams further proposes that supporting a locally focused economy begins with the need of building community within the workplace and connecting to communities where these businesses are located. What is a way that this can this be fostered? One way is via employee ownership of businesses, creating the kind of place presence yielding myriad benefits, both within a business and beyond to the larger community. This approach is presented in the next section.

## The Local Ownership Model[2]

Employee ownership is a way to strengthen ties between business and community, as it tends to blur the lines of how businesses operate, making them more similar to a social enterprise in many aspects more socially responsive as mentioned in the discussion about the fourth sector in a prior section. When employees are empowered as owners, it builds capacity spilling over to the community where employee owned enterprises are located. Abrams, in working with his own employee owned enterprise finds that they attempt to be socially purposeful,

> "by using the financial resources and the web of relationships that derive from our work to help solve community problems and to encourage a better future for the place we live and work. We bring an entrepreneurial approach to these efforts, taking risk and learning from both our public failures and small successes…this is the place we know best…and (we are) doing everything that we can to make a difference in the quality of our community and our economy" (Abrams 2008: 35).

This commitment to place and community should be regarded as an invaluable asset for host communities as well as for supporting positive community economic development initiatives.

---

[2] This section was excerpted with permission from the authors from Phillips et al. (2013).

Employee ownership of businesses is gaining recognition and interest across the US and the world. There are myriad advantages for employees, not only for feeling empowered and part of the company, but also because employee owned businesses tend to provide better pay and benefits, including for retirement. The national organization, Ownership for All, points out the following benefits of an employee ownership business model:

1. Preserve jobs and local ownership. Rather than closing up shop or selling to a competitor, small business owners can sell to their employees. This roots the business and its jobs in the community, and provides a way for employees to share in the wealth created by the business.
2. Provides an exit strategy for owners to preserve continuity in the business. Most owners eventually want to leave the business. Selling to the employees is a way for an owner to exit gracefully, handing off responsibility to managers and ownership to employees at a pace that makes sense for all involved.
3. Improves company performance. When employee ownership is combined with participation in decision-making, businesses often see significant increases in performance (Ownership for All 2012).

In Vermont, the Vermont Employee Ownership Center (VEOC) is a nonprofit organization focused on promoting awareness about the value of employee ownership structures by providing information, resources and technical assistance to owners interested in selling their businesses, and to their employees, employee groups interested in purchasing a business, and entrepreneurs seeking to establish new businesses using an employee ownership model. Currently there are between 30 and 40 employee-owned businesses in Vermont. There are several different forms of employee ownership; described here are the two main forms:

1. Employee stock ownership program (ESOP)
   An ESOP is an employee benefit plan investing in stock of the sponsoring company. Employees are "beneficial owners" of company stock through a trust. ESOPs are expensive, but there are significant tax advantages for shareholders, employees and the company. It is estimated that there are about 10,000 employee-owned businesses in the US; more would develop if additional support was in place.
2. Worker Cooperative
   A worker cooperative is a business in which the workers are equal owners and have control of major company decisions. Profits are usually distributed in proportion to the number of hours worked in day-to-day operations (VEOC 2012).

In Burlington, the value of employee owned businesses gained attention in the early 1980s. A Community Advisory Board was formed to look at this issue, the idea of local ownership as a priority was the response. This idea has continued to grow, with locally-focused business as one way to overcome the challenges of maintaining a tax base while serving as a regional economic and service hub.

The idea of building businesses with strong ties to the local culture, and the uniqueness of a place, is an underlying foundation for these efforts.

Burlington chose to guide development, providing the urban amenities for a rural region without gaining the tax revenues of newer growth nodes in suburban locations—this requires a new way of thinking about how to balance economic demands with social and environmental needs. Burlington has recognized that the most direct remedy—in the past and now—is to continue to support existing employers with job training and counseling programs to match resident skills with existing and future job creation. A more long term innovative initiative is that of the Burlington Local Ownership Project, a special initiative of the City designed to encourage the start-up of a range of locally owned and controlled, for-profit business enterprises and nonprofit and civic organizations—the "social enterprises" needed to foster well-being and build resiliency.

They created a long-term economic development framework focusing on local ownership with a preference for employee ownership. The overarching economic development approach focuses on the concept of locally owned businesses—fusing local business opportunity with employee development. Nearly 30 years later, Burlington is still following this overarching economic development framework, because of a firm belief this supports and fosters a strong local economy.

The choice of a local ownership oriented business development strategy, from a community economic development perspective, is based on the following assumptions:

1. That successful, locally owned businesses will, over the long term, provide more stable employment opportunities for Burlington residents since key corporate decisions will tend to be made by residents with a long-term interest in the future health of the Burlington economy.
2. That successful, locally owned businesses will strengthen the local Burlington economy as both wages and profits are more likely to be retained and reinvested by local owner/employees.
3. That successful, locally owned businesses, being more familiar with local resources and institutions, are more likely to hire, train and promote local residents, therefore promoting a higher percentage of quality job opportunities for Burlington citizens (Phillips et al. forthcoming).

Further, the City of Burlington states that employee owned and controlled businesses should be particularly encouraged because of the following characteristics and anticipated impacts:

- Their demonstrated performance potential. Studies have found employee owned businesses outperform conventionally owned business structures on measures of productivity and profitability;
- The breadth of local ownership which they can provide—in placing long-term strategic decisions affecting the local economies in the hands of a broader number of local actors rather than one or two local entrepreneurs;

- The quality of the employment environment they can create by involving local residents in decisions which affect companies that they own; and
- The fundamental equity and fairness of employee ownership as a business structure—which helps distribute the gains of economic success to the people most responsible for that success—the blue, white and green collar employees (all levels of skills and responsibilities) working under the same roof together.

Within this local ownership business development strategy, a number of enterprise structures are supported including: (1) Conventional, individually owned, entrepreneurial start-ups—possibly taking advantage of city sponsored incubator business space; (2) Conventional, family owned entrepreneurial start-ups; and (3) Employee owner and controlled businesses.

## Implementation

Actual strategies for encouraging local ownership of business can be summarized under three basic categories:

1. Improve substitution start-ups where new, locally owned business is started to produce a product or a service that major local employers must presently import from out of state or beyond. Major employers stand to benefit from these enterprises through the provision of a ready, convenient source of supply reducing their need to carry excess inventory. New, locally owned enterprises would receive a temporary, sheltered, local market which can assist a company in its early start-up stages.
2. Conventional, entrepreneurial start-ups where local entrepreneurs proceed to organize a locally owned business on the strength of a new product or service idea designed for a variety of markets—local, national and international. Likely sources of new business ideas could include university research and development centers, oriented towards encouraging local ownership.
3. Conversions of retiring owner businesses where existing healthy, local businesses find no likely or desirable conventional outside buyer and where local, internal management or management/employee groups move to purchase the firm themselves (Phillips et al. 2013).

Various financing and technical assistance structures exist to help entrepreneurs interested in developing locally owned businesses, including the Burlington Revolving Loan Program, several socially conscious investment funds, the Cooperative Fund of New England, the ICA Group for community lending, and the Vermont Employee Ownership Center.

An example of an employee owned business, and "For Benefit Community Business" is Gardener's Supply, a company which has benefited from the local ownership approach and which in turn has served Burlington and enhanced its community well-being. The City of Burlington began working with them nearly 30 years ago when they were still a small start-up company. They have since grown tenfold and as of December 2009 sold 100 % of the company to their workers. Burlington's and Vermont's creative and independent nature helped fuel a decision for Gardener's Supply to invest in their employees. In 1987, after their third year in business, an Employee Stock Ownership Program (ESOP) was adopted allowing all employees to earn stock and share in company profits. Employees are encouraged to learn about the entire dimension of the business, empowering them to actively participate in guiding current practices and future outcomes. They found that by staff serving as owners, creativity and commitment needed to be responsive to customers is enhanced, as well as sustaining a vibrant focus on gardening and cultivating a compassionate corporate culture in which all are rewarded, including the community in which they operate.

The emphasis of developing a gardening centered company reflects the area's valuing of their natural landscape and rich agriculture heritage, much of which is characterized by sustainable practices. Gardener's Supply was also instrumental in reclaiming floodplain areas at its headquarters site that is now a thriving urban agriculture resource called Intervale. The program includes farm incubation as well as education about resource management. It is recognized as a national model for preservation and urban agriculture and encompasses hundreds of acres. It is a reflection of how a vested community business can positively influence community well-being for its host area on a variety of fronts—economic, social and environmental.

## Conclusion

There are numerous ways to approach development of a locally focused economy, not the least of which is to provide a supporting policy and governance structure. As noted, there are alternative organizational structures emerging to foster socially responsive businesses as well as socially focused organizations. An indicator framework has been suggested with ideas to gauge progress towards a locally focused economy, and the case of Burlington, Vermont is presented to illustrate how one community has fostered development by supporting locally focused enterprise. Because these activities help increase social capacity and in turn generate positive community development outcomes, it is proposed that community well-being is enhanced. Burlington is often listed as among the top communities in the US with high quality of life and desirable places to live, with its locally focused economy as one of the major reasons. This is turn leads to a greater sense of community well-being as part of a thriving culture with a strong sense of presence and vitality.

What are implications for others interested in fostering community well-being via the avenue of locally-focused, community businesses and enterprises? There are several aspects to first consider, including the orientation of local governance—is there willingness to help foster an environment for supporting entrepreneurship and social enterprise? Can policies be changed or created to provide tangible support?

These policies include those for fostering ownership as described in prior sections of this chapter, financial programs for encouraging community lending and investment, and mentoring and business development efforts including business incubation and shared marketing support. Another aspect is to consider the market —what is the culture of the local area? Are residents (and visitors) willing to support community business and enterprise, and are there opportunities for developing a strong presence, such as a focus on local foods production, for example? Starting with an assessment of existing resources and policies for supporting entrepreneurship will provide a basis for gauging additional policy and support changes needed.

Community business and social enterprise aid in creating a definitive sense of place with connections and relationships strengthened, and by so doing, can help an area thrive with positive community development and community economic development approaches. Fostering and supporting this type environment goes far beyond the quantitative aspects of an economy, and touches on many aspects central to community well-being reflected in the framework of this book.

## References

Abrams, J. (2008). *Companies we keep, employee ownership and the business of community and place*. White River Junction, VT: Chelsea Green.

Local First Vermont. (2012). Downloaded April 25, 2012, from www.vbsr.org.

Ownership for All. (2012). Downloaded May 1, 2012, from http://ownershipforall.wordpress.com.

Phillips, R., & Pittman, R. (2009). *Introduction to community development*. London: Routledge.

Phillips, R., Seifer, B., & Antzcak, E. (2013). *Sustainable communities: Creating a durable local economy*. London: Routledge.

Sabeti, H. (2009). *The emerging fourth sector*. Washington, DC: The Aspen Institute.

Schuman, M. (2007). *The small-mart revolution, how local businesses are beating the global competition* (pp. 46–50). San Francisco: Berrett-Koehler Publishers Inc.

Vermont Businesses for Social Responsibility. (2012). Downloaded May 1, 2012, from http://vbsr.org/index.php/pages/static/cat/about_vbsr/. Vermont Employee Ownership Council.

Vermont Employee Ownership Center (VEOC). (2012). Accessed Jan 13. www.veoc.org

# Community Bonding and Community Well-Being: Perspectives from a Community Development Council in Singapore

**Leng Leng Thang, Seung Jong Lee and Youngwha Kee**

## Introduction

Among the losses confronting societies experiencing rapid economic development, urbanization and industrialization since the 1970s has been a sense of loss of community. Chui (2004), in referring to the developments in Hong Kong, has argued that the 'community eclipse' or community decline which happened in Hong Kong is more than a natural occurrence due to urbanization as theorized by Stein (1960) and others. Instead, various factors such as economic, socio-cultural and government policies play a role leading to the demise of community sentiments. In particular, urban sprawl as a result of government public housing and new town policies have uprooted residents and disrupted pre-existent communities.

In Singapore, similar public housing policies in mass scale—while well recognized for its effectiveness in meeting serious housing shortage—are also said to have caused the loss of 'warmth, personal touch, and connections of our old kampongs (village)'. As residents moved from closely knitted community where 'neighbors know each other by first name, help each other when the need arose, live and play together' to high rise public housing with modern amenities, interaction dwindled within the community 'as residents now live behind rod iron gates' (Rasheed 2007).

An inquiry about how people in Western Australia felt about their society, such as what they saw as missing and would like to see in the future has found that

L.L. Thang
National University of Singapore, Singapore, Singapore

S.J. Lee (✉)
Seoul National University, Seoul, Republic of Korea
e-mail: slee@snu.ac.kr

Y. Kee
Soongsil University, Seoul, South Korea

 
S.J. Lee et al. (eds.), *Community Well-Being and Community Development*,
SpringerBriefs in Well-Being and Quality of Life Research,
DOI 10.1007/978-3-319-12421-6_4

people felt strongly about the 'loss of community' or 'loss of identity' in modern society, and rebuilding community structures was high on the priority list for the future among the people.

The village community is often nostalgically interpreted as an ideal of human relationships and community bonding. However, such gemeinschaft type of relationships is also cautioned as largely an ideal which could be oppressive in reality. According to Tonnies who coined the contrastive typology of gemeinschaft/gesellschaft (usually translated as community/society), these are ideal types which never existed. He defines gemeinschaft-like relations as based on natural will with key elements such as sentiment, tradition and common bonds. It is characterized by a "strong identification with the community, emotionalism, traditionalism and holistic conceptions of other members of the community (i.e. viewing another as a total person rather than only as a segment of his status or viewing a person as signification in her own rights rather than as a means to an end) (Lyon 1989: 7). In contrast, gesellschaft-like relations are based on a rational will governed by rationality, individualism and emotional disengagement. It is characterized by "little or no identification with the community, affective neutrality, legalism, and segmental conceptions of other members of the community (ibid.). Instead of expecting human organizations to fall into either one of the two types, Tonnies maintains that it is more realistic to expect them to fall somewhere between the two, "what they represent are ideal types, and they serve as standards by which reality may be recognized and described" (Lyon 1989: 8).

Since Tonnies' seminary work—hailed as possibly marking the beginning of community sociology, much of the concerns in works relating to community has surrounded around the theme of "the loss of gemeinschaft-type relationships in an increasingly gesellschaft-dominated society" (Lyon 1989: 8).

In Singapore, the loss of community spirit is well recognized as an inevitable process of development. Along with nostalgic remembrance of the gemeinschaft-like 'good old kampong days', there are nonetheless pragmatic realization of progress achieved through urban redevelopment. As Chan Soo Sen, the then parliamentary secretary of Ministry of Community Development commented, "none of us want to return to the old days of poverty and deprivation just to achieve social cohesion". However, realizing that at the same time, gesellschaft-like social relationships will be detrimental to social harmony and hence nation-building pertinent for a nation who achieved independence only in 1965, he strategically argues for a revival of gemeinschaft-like relationships, as "the community spirit from the kampong days may hold the key to our effort of building Singapore into our best home" (ibid.: 93).

The Singapore government is probably one of the most proactive governments around committed to shaping community life. The wide array of government-related community and grassroots organizations such as community centers and resident organizations dotting the local constituencies have without doubt contributed to civic life; although the co-optive process, such as the appointments of grassroots leaders by the government has invited criticisms questioning its intentions (Vasoo 1994; Ooi and Koh 2002).

The local administrative unit known as Community Development Council (CDC) set up nationally since 1997 is another new addition to the network of government-initiated attempts to foster community bonding and better the lives of its people. Although the same co-optive approach has again drawn skepticism, from the viewpoint of its contribution to civil society, CDCs efforts to seek ways to encourage citizens to play their roles as volunteers and active community members so as to nurture self-reliant, self-governing local communities are "promise of building a strong and vibrant civil society alongside a strong and effective State" (Ooi and Koh 2002: 100).

This chapter, in focusing on community well-being as a conceptual idea, has chosen to focus on examining the role of CDCs in enhancing community well-being in the context of Singapore. Tasked to find ways to bring people closer, to help those in need, and to make a community a caring one (Rasheed 2007), the CDC model provides an interesting 'experiment' on finding ways to revive/recreate gemeinschaft-type relationships in an increasingly gesellschaft-dominated society. How do the developments inform us on the conceptualization of community well-being? What are some challenges that may limit CDCs' capacity in contributing to community wellness?

In the following, the chapter will first provide a brief note on the concept of community well-being, following which a background overview of developments of CDC in the context of Singapore's efforts in community bonding is presented before focusing on one of the five CDCs, the Central CDC as a case of discussion.

## A Brief Note on Conceptualizing 'Community Well-Being'

As noted in an earlier of this volume, a literature review on the discourse of community well-being has revealed varying definitions and discussions with related concepts such as the UK sustainability Indicators and Bhutan's 'Gross National Happiness Indicators'. Taking into consideration the broad sphere of concerns and priorities, they have subsequently proposed a framework to classify the different related concepts into four categories, where 'community well-being' would belong to the fourth category being the most comprehensive, and analyzing at the community as opposed to the individual level. The proposed framework based its foundation on a multi-dimensional approach of resources and capital. In this respect, it parallels Communities Indicators Victoria (CIV) established in 2007. CIV highlights the importance of recognizing community well-being as encompassing economic, social, environmental, cultural and governance goals. Within the comprehensive coverage of CIV, the concerns with the loss of community may seem to be limited within the domain of 'democratic and engaged communities'. Under the new proposed framework, the concerns could be further captured under two capital domains, the 'social capital' domain which includes trust, commitment, community bonding, participation; and the 'cultural capital' domain which include leisure, sharing and mutual help.

However, as Cox et al. remind us, "the highest priority for CIV and other similar initiatives —is to test and understand the extent to which the availability of comprehensive local community well-being indicator data does indeed improve democratic citizen engagement and policy outcomes" (p. 79). This chapter, in focusing on the role of Singapore's Community Development Councils in promoting community bonding thus focus on the aspect of community well-being which emphasizes the building of social capital for sustainable and resilient communities, one where people live together in harmony and with satisfaction.

Viewed from the perspective of Tonnies' typology, a community that has achieved community well-being would also have somewhat achieved gemeinschaft-like characters with strong community and bonding among its people. In the discussion of CDC below, community well-being will be discussed primarily from the software aspect of community bonding.

## The Need for Community Bonding and Community Development Councils

As mentioned earlier, the Singapore government plays an important role in fostering community bonding. Communal harmony and social cohesion have always been of high priority in nation-building and political stability of the state. Faced with the challenging task of creating a national identity from a disparate, diverse, multi-racial and multi-religious society where community bonding during the colonial period more often than not meant that each ethnic group looked out for themselves through their own developed "self sufficient" communities and groups, community bonding for the government under the People's Action Party (PAP) is inevitably linked closely with inter-racial solidarity (Yong 2004). These initiatives are usually related to People's Association (PA), a statutory organization formed since 1960 with the Prime Minister as the chair of its management board to provide centralized direction aligned with the government over the community-based organizations (Seah 1973). Over the years, the community-based organizations accountable to PA have expanded from community centres and Citizens' Consultative Committees in the 1960s, to Residents' Committees in the 1970s and Community Development Councils in the 1990s.

The rapid relocation of people into public housing estates since the 1960s—usually high rise apartments disconnected from their familiar neighborhoods—further accentuated the need to promote better neighborhood cohesion and integration among its people. Singapore faced dire housing shortage by late 1950s, with only 40,000 units available through the Singapore Improvement Trust (public housing sector) and private sector between 1947 and 1959 although the population has reached beyond 1.5 million at that time. The problem was resolved with the Housing and Development Board (HDB)'s successful building programs which began since 1961. By 1965, 10,000 units of flats were built. In 10 years, HDB have provided home for about 30 % of the population, and by 1989, it has expanded to cover above 75 % of Singapore's population (Vasoo 1994). Today, HDB houses more than 80 % of the population of

5.18 million. With more than 90 % owning their own flats, the home ownership scheme introduced since 1964 has enhanced one's financial security as property owners, as well as one's commitment to the place and the nation. HDB also imposes a quota system for different races since 1989 to prevent ethnic ghettos, foster racial tolerance and racial integration. Singapore has three major races, of which Chinese comprises about 75 %, Malay about 13 % and Indian 9 %.

However, while the high rise blocks have changed Singapore's landscape and effected social transformation as people parted with old style kampong housing, slums and squatter living to embrace new modern lifestyle in the public housing estates that came with more facilities and better hygiene, the mass relocation have also resulted in the feeling of the loss of community. Such sentiment was expressed through an older HDB dweller as follows:

> In the kampong, everyone knew each other. There was no need to shut your doors the whole day. If a stranger came to the kampong, we would inform each other and strangers rarely came in the night. I wish I could travel back in time and return to the kampong lifestyle (National Archives 1993: 83).

Fully aware of the problems of estrangement and community disintegration which may face HDB dwellers, and thereby challenge the national objectives of maintaining racial harmony and others, HDB has placed the building of cohesive communities as one of its key priorities. The provision of community spaces in the housing precincts, between the blocks and the void decks on the ground floor of the blocks are such spaces encouraging interaction and acceptance of diversity. The first page of an HDB publication titled "Homes: 50 years of housing a nation" (Fernandez 2011) has articulated "The HDB Experience" as follows:

> Rather than just building blocks of flats, HDB has strived to build communities. Its plans and designs have incorporated facilities and spaces for residents to mix and mingle, and forge ties. These enable them to relax to the sound of birdsong, enjoy a game of chess or basketball, or even do some gardening or kite-flying. Ground floor void decks also process a place for all sorts of major life events, from weddings to funerals.

In the recent years, HDB has also proactively organized various activities to foster community bonding among the residents within the same vicinity, such as welcome parties for new residents in newly completed blocks.

In the vastly HDB housing environment, besides the HDB bonding activities, the network of state-initiated grassroots and para-political organizations are the various welfare and self-help oriented organizations involved in community services. Many of them are housed at the void decks. Referred to as civil society organizations (CSOs) by Ooi and Koh (2002), these non-governmental organizations [also commonly called voluntary welfare organizations (VWOs)] such as family service centers and neighborhood link centers may be co-funded and received administrative support from the government. The CSOs and the state-initiated organizations generally take the 'welfare approach' providing direct services to help the disadvantaged and the poor (Ooi and Koh 2002).

There is indeed a wide array of community-based organizations nationwide set out to connect the people with their community and service needs, as well as to connect the state with the people. However, the structure and organization of the state-initiated grassroots and para-political organizations, in particular, are often critiqued for its close link with the PAP party, causing concerns over whether the interest of community residents are compromised over political desire (Vasoo 1994; Ooi and Koh 2002). Among which, CDCs, the relatively new comer of the state-initiated community-based organization has also been a subject of critique.

The Community Development Councils were set up in 1997 as part of the local governance structure devoted to developing the software aspect of promoting racial harmony, strengthening social cohesion and strengthening community bonding. Each CDC is managed by a mayor appointed by the chairman or deputy chairman of the People's Association Board of Management, supported by district councilors, resource panel or committees who are volunteers and also appointed, and paid staff from General Manager to other managers and staff. Active citizens in the community form the volunteers at the base of the structure. Each CDC receives an annual resident grant of $1 per resident living in its District to fund its programs. CDCs are encouraged to raise their own funds of their programs with three of four times of matching grants from the government for each dollar raised. The operation costs of CDC offices are funded by the government. Each CDC also receives fund from the government to manage welfare programs such as public assistance and Medifund (financing for medical expenses to the needy).

Thio (2009) has referred to CDCs (and Town Councils) as local government in a muted form due to its connection with partisan politics, such as the appointment of mayors from the ruling party's members of parliament instead of running local elections (George 2000; Thio 2009). In a way, the strong backing has ensured the success of CDCs tasked in its vision to build a vibrant community through the strategic tasks of ABC—Assisting the needy, Bonding the people and Connecting the community so as to build a great home and a caring community (CDC Annual Report 2010).

In evaluating the impact of CDCs, George (2000) contends that "perhaps their biggest impact on civil society is providing a mechanism for cooperation between government grassroots activists and non-government organizations" (p. 153). Through CDCs' funding resources and initiatives, grassroots organizations are coming together more with the NGOs for various community projects. They are complementary matches, while grassroots organizations have an understanding of local community needs, the NGOs have professional expertise and experiences to meet the needs. In fact, working together with the stakeholders and partners within the community has been an important strategy for the CDCs towards the creation of community and the fostering of community spirit. The discussion in the next section focusing on the Central CDC will provide a more detailed understanding of the strategies of community bonding in the CDCs.

## Central CDC and the Bonding of Community

Central CDC is the largest CDC among the five CDCs in the country, serving about 1 million residents in the district. The current mayor of the Central CDC is Mr. Sam Tan, heading a current staffing of about 170 people.

The Central part of Singapore is a mix of old and new Singapore, where modern Central Business District and the shopping belt lie adjacent to the older historic areas with distinct ethnic flavors such as Chinatown and Little India. The central part of Singapore is also where early public housing projects first started in the 1960s and 1970s, thus the Central Singapore District is characterized with older residents from the mature housing estates. With two-thirds of the low income rental flats situated in the district, it has the highest proportion of low-income residents among the five districts. However, at the same time, Central Singapore houses some of the most affluent population in its pockets of expensive residential areas. As the demographics of each district determine the types of programs initiated for the residents, it is expected for Central CDC to tend to have more programs for the seniors, while others, such as the Northeast CDC located in the highly residential areas more populated with younger residents tend to initiate more training programs for younger residents.

## Programs at the Central CDC and Community Bonding

In the ABC of the strategic thrusts of the CDCs, the welfare-focus of 'assisting the needy' seems at first glance to be quite unrelated from the other two thrusts of 'bonding the people' and 'connecting the community' which have direct reference to community bonding. Nevertheless, social assistance work is an integral part of community bonding efforts, as "it would be meaningless if we championed social harmony when some people go hungry on empty stomachs" (Rasheed 2007).

CDC acts as a one stop referral and help center for needy residents. In Central CDC, besides the office at the CDC, a new satellite office located at a community centre has also been set up to offer easy access to residents, and if residents in need have difficulty coming into the office, officers can be arranged to pay home visits to offer assistance with application. Besides a variety of government programs and schemes administered by the CDC to offer direct help, CDCs also initiate their own support programs as well as provide referrals to other government agencies, NGOs, grassroots organizations and self-help groups. In fiscal year 2010–2011, Central CDC is reported to have assisted 16,016 residents under the various national social assistance schemes and provided financial help to 1,090 residents through its local assistance programs such as temporary relief schemes and disbursements from various charitable foundations (Annual report of Central CDC 2012).

It should be noted that since the government's establishment of the Community Care Endowment Fund (ComCare Fund) in 2005, national social assistance schemes are now known as national Comcare schemes where they are administered by CDCs for their respective residents. ComCare is guided by the principal of promoting self-reliance, therefore besides providing financial help, such as Public Assistance to those qualified to receive monthly welfare assistance from the government, financial assistance for childcare, kindergarten and student care fee assistance for children of low-income families, it aims at coordinated efforts to enable families to become self-reliant eventually. For example, self-reliance and self-help are promoted through programs and schemes to help individuals and families to obtain better education and employment opportunities.

Employment assistance to residents is another important program under the banner of 'assisting the needy'. The employment services provided by Central CDC include recruitment events such as free workshops on employability skills and walk-in interviews with employers held in the district. Recently, the Central CDC has consolidated the information necessary for employment by revamping the employment portal to include links of online job portals, information on the training opportunities available from various agencies in Singapore besides employment-related information from Central CDC.

Over the years, besides the national assistance schemes and employment services, CDCs have also come forward with their own local schemes to assist the needy in their district. In the Central CDC, a savings program called the C.A.S.H (Cultivate A Savings Habit) program sponsored by Maybank Singapore have been implemented to encourage low-income families to save in order to improve their financial situation. The nine-month savings scheme piloted in April 2011 for families with monthly household income of S$1,800 or less requires participants to attend the one-day Talking Dollar and Sense workshop organized by Central CDC about managing finances. They will then follow up with deposits in their savings accounts at least once every three months, where Maybank Singapore will match the amount they save to a sum of S$1000. A write-up about the program reported more than 170 participants in the program where they found the workshop useful in teaching them how to budget for their needs and wants with what they have (Huang 2012).

## *Bonding the People and Community Through Actions*

With the vision of "an inclusive, vibrant and self-reliant Central Singapore Community", Central CDC organizes various programs to cater to different causes and age groups. These community programs are generally classified under 'community bonding' and 'community services'.

Programs under 'community bonding' are as follows (refer to Appendix A for details):

- Arts programs
- Environment Programs
- Racial Harmony Programs
- Sports and Health Lifestyle Programs

The following are programs under 'community service' (refer to Appendix B for details):

- Elderly Programs
- Financial Literacy Programs
- Youth Programs
- Pass it On
- Project Include
- Mayor's Imagine Fund
- Social Enterprise Fund

As Appendix A and B show, there are several projects or schemes under each of these categories. The variety of programs show creativity in efforts to bond the people and the community, as well as the CDC's constant look out for ideas and the flexibility of adapting from different successful community ideas locally and internationally in community bonding experiments. For example, the Mayor's Imagine Fund was adapted from the Imagine Chicago community initiative originated in Chicago. The Orange Ribbon Celebrations (ORC) under the racial harmony program came from the Orange Ribbon idea adopted by the United Nations to mark the International Day for the Elimination of Racial Discrimination. The Central CDC initiated the first Orange Ribbon Celebrations in July 2006 to promote the understanding and appreciation of Singapore's ethnic and cultural heritage to residents; and the success of the ORC has prompted PA to enlarge the idea to a national-level racial harmony celebration in 2008.

Among the various Central CDC programs, Community Life Arts Programs (CLAP!) is a flagship program started since 2001 to organize regular-arts-based community outreach program. In 2012, the CLAP! Program further receives increased publicity by collaborating with the Esplanade—theatres on the Bay to bring quality arts performances to different venues within the district every month of 2012.

In the recent years, CLAP! has expanded at the suggestion of grassroots organizations, where Central CDC begin to 'francaise' it out by providing seed funding and support for grassroots organizations to run their own arts events for their communities. Funding from CDC has become one channel encouraging active citizens' participation in organizing community projects of different nature.

Under "community services", the 'Bright Homes' scheme started in 2006 to help lonely elderly living in the district has also developed into a funding program where volunteer groups and community partners may seek funding as they organize and plan assistance to meet the needs of low-income senior citizens living in one to two room rental flats. To encourage befriending and regular contacts with the elderly, the scheme has set the condition requiring the volunteers to commit to organize their sessions once a month for a minimum of six months. Volunteer groups for the program has come from various organizations such as schools, companies and grassroots organizations and activities they have organized included home cleaning, parties and excursions. There were 18 Bright Homes program in 2010–2011 benefiting more than 900 elderly with the engagement of 360 volunteers (Central CDC Annual Report 2012).

In April 2012, the Central CDC has launched a new three year project called "Hands for Homes program" requiring more than S$200,000 of funding each year. This program began with concerns from grassroots leaders in the older area of Kreta Ayer with the bedbug-infested mattresses and hygiene of needy elderly living in rental flats in their area. With a lack of funding and volunteers to carry out the project, they approached the Central CDC for assistance, in which it was developed into a new program where the Central CDC plays important roles in providing funding, locating sponsors and volunteers. The program aims to provide a more comprehensive outreach to the needy elderly, besides the spring cleaning of beg-bug homes and the provision of new mattresses, there will also be other services to enhance their physical and social well-being, including the provision of anti-slip floor mats and induction cookers for home and kitchen safety, the installation of energy-saving light bulbs to reduce energy consumption, and social activities for the elderly while their homes are being cleaned. It is estimated that 400 volunteers are needed to spring clean the homes of 200 elderly affected by bedbugs. The project is costly due to the cost of engaging pest control services to disinfect each home; the progress is also slow as spring cleaning the homes is a labor-intensive effort. As a one-day program of a session of spring cleaning carried out in mid May 2012 with corporate volunteers joined by the Member of Parliament of the area and the Mayor of the Central CDC show, it takes more than 20 volunteers to clean up 10 homes in a day. Despite the challenges, the program has nevertheless contribute to a sense of community spirit and mutual help, as it provides an opportunity for the elderly residents to leave their homes to interact with each other and with the volunteers during the activity.

From the current programs and projects available at the Central CDC, we can notice that there are lesser number of large scale events compared with the smaller but more regular programs. This represents a shift in approach to more regular, sustainable programs which will be more effective in building bonding among the people than large scale events (such as carnivals) which attract a large crowd and good publicity but may not be as effective in encouraging spontaneous bonding. The change in approach also implies a need to change the evaluation of the impact of CDCs on community bonding. For example, the process should be regarded as equally important as the outcome in promoting community bonding. The measure of sustainability of a project inevitably relates to the depth of social capital, the friendship fostered in the process and the engagements that allowed for the fostering of a sense of community and belonging.

## Conclusion

Several strategies to promote community bonding and the enhancement of community well-being are provided in this section. While lamenting that the fast-pace Singapore society has lost a sense of community and the 'kampong spirit', the establishment of CDCs tasked with bonding the community has shown their efforts to create/revive/recreate/re-engage and at times to re-define the community. The discussion of the socio-political background leading to the need for community bonding, and the various sustainable projects and programs organized or supported by the CDCs serve as concrete case evidence of the significance of social capital and cultural capital in fostering community well-being. To understand what constitutes community well-being, it is indeed important to have knowledge of the cultural and socio-political context that defines the society that the community belongs to. In addition, we can also argue that by providing funding to kick start ground-up community initiatives, CDCs are also involved with enhancing the economic capital necessary for community development.

In general, we can summarize the strategic roles of CDCs in enhancing community well-being as follows: the connector and the community venture capitalist. The 2010–2011 annual report of the Central CDC has used the image of jigsaw puzzle to symbolize their role—like bringing the different pieces of puzzle together to complete the picture, CDC has a role in connecting the community players such as the state-initiated grassroots with the NGOs, and bringing together corporate and individual sponsors and volunteers, school volunteers for a social cause and eventually to build a socially coherent and vibrant community. The newly set up "Hands for Homes Program" by the Central CDC is a succinct example of the jigsaw puzzle image, where the Central CDC comes together with grassroots organization, as well as connecting the corporate volunteers and sponsors to specifically promote the well-being of the needy elderly.

The analogy of the CDC as providing community venture capital was mentioned during a conversation with the General Manager of Central CDC. This aptly describes the wide array of funding programs operated by the Central CDC to encourage ground-up activities and projects. Many of these projects are small scale, for example, the Healthy Lifestyle Clubs (HLCs) can be formed by any group (including grassroots organizations, schools, community groups etc.) with a minimum of 15 members for regular exercise and sports to foster a healthy lifestyle and interactions. There are currently 137 HLCs with 13,000 members.

In addition, the CDCs' role in promoting community bonding should also be considered in the Singapore multi-racial environment of maintaining racial harmony. Such an objective forms the underlying basis of community bonding in the Singapore context, where the evaluation of the extent of a CDC's success in fostering community bonding would inevitably include their extent of engagement among the different races. This characteristic highlights the need for attention to local uniqueness and variations in understanding what constitutes community well-being.

Thus, has the presence of CDC help in concrete ways to promote community bonding, and thereby community well-being in Singapore? In its efforts to foster the well-being of individuals through the meeting of their needs, and by offering opportunities through funding support and from its role as the connector to enable an idea conceived on the ground to lift off so as to benefit the community and enhance the well-being of both the individuals and the community they live in, the CDC concept has certainly created/revived/re-create/strengthen the community—whether conceived as a large entity of the area of the boundaries of a CDC or a small precinct.

However, challenges to community well-being still remain. For example, the critique from political perspective about community and partisan politics, and whether co-optation as seen by some will eventually lead to expanded political outreach than civil society expansion will continue to be an issue of contention in impacting the sense of optimal community development (George 2000; Thio 2009; Ooi and Koh 2002). There is also the challenge of increasing awareness and reaching out to more people in encouraging active citizenry through volunteering and commitment to better the community. George (2000) has questioned if the CDC may tend to attract volunteers who are already from the grassroots organizations and NGO activists instead of expanding to include more fresh volunteers among the residents. The issue of engaging younger people known to be less interested in local community building and bonding in this globalized era is another challenge (Tan 2011). Furthermore, Singapore—as in many big cities, is witnessing an increasingly metropolitan and international influx of people, which sometimes resulted in conflicts with the locals as a result of a lack of interaction. Eventually, community development must address such issues of engaging and integrating beyond those who are born and bred on the land to include foreigners living and working side-by-side so as to achieve community well-being in the real sense.

**Acknowledgments** The authors would like to thank the Central Community Development Council of Singapore. For their kind assistance in the information gathering process.

## Appendix A

See Table 1.

**Table 1** Bonding the people and community: community programs at the Central CDC

| | | | |
|---|---|---|---|
| Arts programs | a. CLAP! community life arts program | CLAP! started in 2001 and aims to bring the arts and performances to residents in the Central District and to promote bonding in the community | In FY 2010–2011, 54 sessions of CLAP! were organized. Together with CLAP! Franchise, the arts programs reached out to more than 50,000 residents in the year |
| | b. CLAP! franchise | CLAP! organized by Grassroots Organizations in their own neighborhood with seed funding and some support from the Central CDC | Currently, there are 12 CLAP! franchises in the Central Singapore District. In FY 2010–2011, 98 sessions were organized |

(continued)

**Table 1** (continued)

| | | | |
|---|---|---|---|
| Environment programs | a. Environment fund | Fund given to Grassroots organizations to organize public health and environment-related programs within their constituencies | Examples of projects supported by this Fund: exhibitions on dengue prevention, anti-littering campaigns, community recycling events, art competitions featuring the Go Green message and formation of gardening clubs |
| | b. Project EARTH (Every Act of Recycling Trash Helps) | The project launched in 2009 aims to encourage recycling and is organized in partnership with the National Environmental Agency (NEA) Central Regional Office | This project is part of "Mayor's Green Challenge" series in the 10-year Central Singapore Environmental Sustainability Plan (ESP). To date, it has exceeded the target of 10,000 tonnes (10,000, 000 kg) with more than 14,000 tonnes of trash has been collected for recycling by partner organisations like schools, shopping malls and corporate companies |
| Racial harmony programs | a. TRUST (The Racial and Religious Unity Steering Committee in Central Singapore) home program | A home visit program aimed at promoting interaction between people of different ethnic and religious backgrounds | Participants will be hosted to a visit by host families who will share information on their daily customs and practices. Host families include grassroots leaders, community volunteers and District councilors |
| | b. TRUST calendar | Special calendar in four languages that carries the dates of the major ethnic and religious festivals. It is produced by Central CDC every year (since 2004) | The calendar is designed around themes which reflect the material culture and customs of our ethnic communities |
| | c. Orange Ribbon Celebrations (ORC) | To promote the understanding and | First launched in 2006, the success of |

(continued)

**Table 1** (continued)

| | | | |
|---|---|---|---|
| | | appreciation of Singapore's rich ethnic and cultural heritage to residents | the Central Singapore District ORC has elevate the idea to a national-level racial harmony celebration by PA. In 2008, the ORC was launched as a national initiative spearheaded by OnePeople.sg, together with the 5 CDCs , Self-Help Groups and the Ministry of Community Development, Youth and Sports (MCYS) and Ministry of Education (MOE) |
| Sports and healthy lifestyle programs | a. Healthy lifestyle clubs (HLC) | Launched in 2001 to encourage residents to adopt a healthy lifestyle through regular exercise and sports, and to promote bonding among families and active aging for the elderly | A HLC can be formed by any group (including grassroots organizations, community groups, schools, etc), with a minimum of 15 members. There are currently 137 HLCs with 13,000 members |
| | b. Free kicks program | The program launched in 2003 is one of Central CDC's key initiatives to promote community bonding among youths and families through soccer | It is opened to boys aged 7–16 years, and girls aged 8–18 years. With a small fee, participants receive 40 sessions of soccer coaching by professional coaches, including coaches from S League Clubs. Youths from low-income families can receive fee waiver The annual highlight of this programme is the Central Singapore Mayor's Challenge Shield, where participants from all the centres come together for a day of matches |

*References* The Central CDC Website (http://www.centralsingaporecdc.org.sg/) and Central CDC Annual Report FY 2010–2011

## Appendix B

See Table 2.

**Table 2** Bonding the people and community: community services at the Central CDC

| | | | |
|---|---|---|---|
| Elderly programs | Bright Homes | Bright Homes was initiated to address the issue of the lonely elderly. Since it started in 2006, it has developed into a funding program for volunteer groups and community partners to assist in meeting the home-based needs of lower-income senior citizens living in 1–2 room rental flats | Bright Homes encourages volunteers to maintain regular contact with the elderly. As a funding condition, they are required to organize a Bright Homes session once a month, for a minimum of 6 months. The volunteer groups range from schools to corporate groups. Activities include conduct homecleaning, parties, excursions, and other activities to engage and bring cheer to the elderly. There are about 900 elderly residents currently engaged by the Bright Homes volunteers |
| Financial literacy programs | StarHub-Central Singapore Nurture Programme | The Nurture Program curriculum aims to develop the literary strengths of children, age 7–12 years old from low-income families, with the belief that education will help the next generation to break out of the poverty cycle | It was started in 2007 by a group of Central Singapore CDC volunteers. In FY 2010–2011, there are 17 nurture centers at community centers and voluntary welfare organizations (NGOs). There are more than 650 children attending the program, taught by 120 regular volunteers |
| Youth programs | High Five Youth (HFY) | A youth volunteer group started in 1999 in Central CDC, it aims to promote a dynamic culture of youth volunteerism which can empower our youths to effect change in the community | The High Fivers have organized several key signature events, including Tapestry (an annual street busking program to raise funds for charitable organizations), D-Act (Action for Dementia) and Travelling Together. They are also involved in the CDC's Bright Homes and Nurture programs |

(continued)

**Table 2** (continued)

| | | | |
|---|---|---|---|
| Pass it on | | A program to help the less fortunate and reducing wastage. It has an online donation portal which allows the public to donate their unwanted but usable household items to the less privileged | The donated items are made available to all Voluntary Welfare Organisations (VWOs) who will match these items to families which needs them |
| Project INCLUDE | | Aan initiative to ensure that all residents, regardless of their disabilities, are able to take part, enjoy themselves, and benefit from communal activities, along with the rest of the community | Efforts by the CDC include institutionalizing practices to ensure that the Central CDC office and programmes are disabled friendly; to engage more agencies in the disability sector to be involved in mainstream programs; and empower others to initiate programs to promote an inclusive Central Singapore District |
| Mayor Imagine Fund | | Mayor's Imagine Fund was launched in April 2002 to promote active citizenry among the residents and to help them realize their ideas for the community | In FY 2010–2011, 39 projects were funded for up to $5,000 |
| Social Enterprise Fund | | The Fund supports sustainable social entrepreneurship projects targeted at benefiting the less advantaged residents of Central Singapore district | The committee will approved up to 80 % of the total project cost, subject to a cap of $30,000. Funding will be reimbursed over a period of 2 years |

*References* The Central CDC Website (http://www.centralsingaporecdc.org.sg/) and Central CDC Annual Report FY 2010–2011

# References

Central Community Development Council. (2012). Annual Report 2010–2011, Singapore.

Chui, E. (2004). Unmasking the 'naturalness' of 'community eclipse': The case of Hong Kong. *Community Development Journal, 38*(2), 151–163.

Community Development Councils. (2010). Annual Report, Singapore.

Fernandez, W. (2011). *Our homes: 50 years of housing a nation*. Singapore: Straits Times Press.

Geroge, C. (2000). *Singapore: The air-conditioned nation*. Singapore: Landmark Books.

Huimin, H. (March/April 2012). *What a bonus! In voices @ Central Singapore*, Issue 63 (pp. 14–15).

Lyon, L. (1989). *The community in urban society*. Lexington, MA: Lexington Books.

National Archives. (1993). *Kampong days: Village life and times in Singapore revisited*. Singapore: National Archives.

Ooi, G. L., & Koh, G. (2002). State-society synergies: New stakes, new partnership. In A. Mahizhnan & Y. L. Tsao (Eds.), *Singapore: Re-engineering success*. Singapore: Times Academic Press.

Rasheed, Z. A. (2007). Speech by Senior Minister of State for Foreign Affairs Mr. Zainul Abidin Rasheed about Community Development Councils in Singapore at the Leadership in Megacities Seminar, Jakarta, 24 May 2007. Available online http://app.mfa.gov.sg/2006/lowRes/press/view_press.asp?post_id=2655. Accessed 17 Apr 2012.

Seah, C. M. (1973). *Community centres in Singapore: Their political involvement*. Singapore: Singapore University Press.

Stein, M. (1960). *The eclipse of community*. Princeton, NJ: Princeton University Press.

Tan, E. S. (2011). Studying community relations. In A. I. Latif (Ed.), *Hearts of resilience: Singapore's community engagement programme*. Singapore: Institute of Southeast Asia Studies.

Thio, L. A. (2009). *Neither fish nor fowl: Town councils, community development councils and the cultivation of local government/governance in Singapore*. http://nus.academia.edu/LiannThio/Chapters/606011/Neither_Fish_nor_Fowl_Town_Councils_Community_Development_Councils_and_the_Cultivation_of_Local_Government_Governance_in_Singapore. Accessed 15 May 2012.

Vasoo, S. (1994). *Neighbourhood participation in community development*. Singapore: Times Academic Press.

Yong, M. C. (2004). Some thoughts on modernization and race relations in the political history of Singapore. In M. C. Yong (Ed.), *Asian traditions and modernization: Perspectives from Singapore*. Singapore: Eastern University Press.

# Index





S.J. Lee et al. (eds.), *Community Well-Being and Community Development*,
SpringerBriefs in Well-Being and Quality of Life Research,
DOI 10.1007/978-3-319-12421-6

CPSIA information can be obtained at www.ICGtesting.com
Printed in the USA
BVOW09s2154151214

379562BV00004B/18/P

9 783319 124209